Lecture Notes in Computer Science 13574

More information about this series at https://link.springer.com/bookseries/558

Wenjian Qin · Nazar Zaki · Fa Zhang · Jia Wu ·
Fan Yang (Eds.)

Computational Mathematics Modeling in Cancer Analysis

First International Workshop, CMMCA 2022
Held in Conjunction with MICCAI 2022
Singapore, September 18, 2022
Proceedings

Editors
Wenjian Qin
Shenzhen Institute of Advanced Technology
Chinese Academy of Sciences
Shenzhen, China

Fa Zhang
Institute of Computing Technology
Chinese Academy of Sciences
Beijing, China

Fan Yang
Stanford University
Palo Alto, CA, USA

Nazar Zaki
United Arab Emirates University
Al Ain, United Arab Emirates

Jia Wu
The University of Texas MD Anderson
Cancer Center
Houston, TX, USA

ISSN 0302-9743   ISSN 1611-3349 (electronic)
Lecture Notes in Computer Science
ISBN 978-3-031-17265-6   ISBN 978-3-031-17266-3 (eBook)
https://doi.org/10.1007/978-3-031-17266-3

This Springer imprint is published by the registered company Springer Nature Switzerland AG
The registered company address is: Gewerbestrasse 11, 6330 Cham, Switzerland

Preface

The 1st Workshop on Computational Mathematics Modeling in Cancer Analysis (CMMCA 2022) was held in conjunction with the 25th International Conference on Medical Image Computing and Computer Assisted Intervention (MICCAI 2022) on September 18, 2022. Due to the COVID-19 pandemic restrictions, CMMCA 2022 was held virtually.

Cancer is a complex and heterogeneous disease that often leads to misdiagnosis and ineffective treatment strategies. Pilot mathematical and computational approaches have been implemented in basic cancer research over the past few decades, such as the emerging concept of digital twins. These methods allow a deeper exploration of cancer from the perspective of computational science, such as the mapping of biological and computational correlations among multiple omics data at various scales and views, in which the multimodal cancer data include, but are not limited to, radiography, pathology, genomics, and proteomics. Motivated by rigorous mathematical theory and biological mechanisms, the advanced computational methods for cancer data analysis are robust and clinically practicable, which will result in strong interpretability in combining clinical data and algorithms in an era of artificial intelligence.

CMMCA provides a platform to bring mathematicians, biomedical engineers, computer scientists, and physicians together to discuss novel mathematical methods of multimodal cancer data analysis. A major focus of CMMCA 2022 was to identify new cutting-edge techniques and their applications in cancer data analysis in response to trends and challenges in theoretical, computational, and applied aspects of mathematics in cancer data analysis.

The workshop attracted worldwide attention, including experts in radiography, pathology, and genomics multi-modality data, as well as learning-based imaging processing and computational modeling for cancer analysis. All submissions underwent rigorous double-blind peer review by at least two members (mostly three members) of the Program Committee, composed of 26 well-known research experts in the field. The paper selection was based on methodological innovation, technical merit(s), relevance, the significance of results, and clarity of presentation. Finally, we received 16 submissions, out of which 15 papers were accepted for presentation at the workshop and inclusion in this Springer LNCS volume.

We are grateful to the Program Committee for dedicating their time to reviewing the submitted papers and giving constructive comments and critiques, to the authors for submitting high-quality papers, to the presenters for excellent presentations, and to all the CMCCA 2022 attendees from all over the world.

September 2022

Wenjian Qin
Nazar Zaki
Fa Zhang
Jia Wu
Fan Yang

Organization

Organizing Chairs

Wenjian Qin	Shenzhen Institute of Advanced Technology, Chinese Academy of Sciences, China
Nazar Zaki	United Arab Emirates University, UAE
Fa Zhang	Institute of Computing Technology, Chinese Academy of Sciences, China
Jia Wu	University of Texas MD Anderson Cancer Center, USA
Fan Yang	Stanford University, USA

Program Committee

Chengyan Wang	Fudan University, China
Eman Showkatian	Med Fanavarn Plus Co., Iran
Erlei Zhang	Northwest A&F University, China
Fang Hou	Massachusetts General Hospital, USA
Harsh Singh	United Arab Emirates University, UAE
Hongmin Cai	South China University of Technology, China
Jin Liu	Central South University, China
Kai Hu	Xiangtan University, China
Liangliang Liu	Henan Agricultural University, China
Mohd Saberi Mohamad	United Arab Emirates University, UAE
Muhammad Waqas	FAST National University of Computer and Emerging Sciences, Pakistan
Rafat Damseh	United Arab Emirates University, UAE
Rizwan Qureshi	City University of Hong Kong, Hong Kong, China
Sherzod Turaev	United Arab Emirates University, UAE
Songhui Diao	Shenzhen Institute of Advanced Technology, Chinese Academy of Sciences, China
Tian Bai	Jilin University, China
Wei Zhao	Beihang University, China
Xiaoying Tang	Southern University of Science and Technology, China
Xing Lu	Zhuhai Sanmed Biotech Ltd., China
Xiu Nie	Union Hospital, Tongji Medical College, Huazhong University of Science and Technology, China

Yaoqin Xie	Shenzhen Institute of Advanced Technology, Chinese Academy of Sciences, China
Yiqun Lin	The Hong Kong University of Science and Technology, Hong Kong, China
Yizheng Chen	Stanford University, USA
Zhaozhao Tang	Shenzhen University, China
Zhenhui Dai	The Second Affiliated Hospital, Guangzhou University of Chinese Medicine, China
Zhiying Xie	University of Washington, USA

Contents

Cellular Architecture on Whole Slide Images Allows the Prediction of Survival in Lung Adenocarcinoma

Pingjun Chen[1], Maliazurina B. Saad[1], Frank R. Rojas[2],
Morteza Salehjahromi[1], Muhammad Aminu[1],
Rukhmini Bandyopadhyay[1], Lingzhi Hong[1,3], Kingsley Ebare[4],
Carmen Behrens[3], Don L. Gibbons[3], Neda Kalhor[4], John V. Heymach[3],
Ignacio I. Wistuba[2], Luisa M. Solis Soto[2], Jianjun Zhang[3,5],
and Jia Wu[1,3(✉)]

[1] Department of Imaging Physics, Division of Diagnostic Imaging,
The University of Texas MD Anderson Cancer Center, Houston, TX 77030, USA
jwu11@mdanderson.org
[2] Department of Translational Molecular Pathology,
The University of Texas MD Anderson Cancer Center, Houston, TX 77030, USA
[3] Department of Thoracic/Head and Neck Medical Oncology,
The University of Texas MD Anderson Cancer Center, Houston, TX 77030, USA
[4] Department of Pathology, The University of Texas MD Anderson Cancer Center,
Houston, TX 77030, USA
[5] Department of Genomic Medicine, The University of Texas MD Anderson Cancer
Center, Houston, TX 77030, USA

Abstract. Pathology is the gold standard for cancer diagnosis. Numerous studies aim to automate the diagnosis based on digital slides, yet its prognostic utilities lack adequate investigation. Besides the inherent difficulties in predicting a patient's prognosis, extracting informative features from gigapixel and heterogeneous whole slide images (WSI) remains an open challenge. We present a computational pipeline that can generate an embedded map to flexibly profile different cell populations' local and global composition and architecture on WSIs. Our approach allows researchers to investigate tumor cells and tumor microenvironment based on these embedded maps of a reasonable size rather than dealing with gigantic WSIs. Here, we applied this pipeline to extract the texture patterns for tumor and immune cell types on the TCGA lung adenocarcinoma dataset. Based on extensive survival modeling, we have demonstrated that by pruning redundant and irrelevant features, the final prediction model has achieved an optimal C-index of 0.70 during testing. Our proof-of-concept study proves that the efficient local-global embedded maps bear valuable information with clinical correlations in lung cancer and potentially in other cancer types, warranting further investigations.

P. Chen, M. B. Saad and F. R. Rojas—Equal Contribution.

W. Qin et al. (Eds.): CMMCA 2022, LNCS 13574, pp. 1–10, 2022.
https://doi.org/10.1007/978-3-031-17266-3_1

Keywords: Cell architecture · Whole slide image · Nuclei classification · Lung adenocarcinoma · Survival analysis

1   Introduction

Pathologic assessments serve as the gold standard for cancer diagnosis. With the adoption of whole slide image (WSI) scanners, tissue slides on glass can be converted to digital format for downstream analysis. Recently, numerous artificial intelligence (AI) studies have sought to automate the tissue level diagnosis [4–7,15,18,22], with some even reported reaching pathologist-level performance [11,19,22]. Compared to cancer diagnosis, there are less attentions paid to predicting treatment response or survival [12,13,21]. Fundamentally, cancer diagnosis is a classical detection or classification problem on the given slides. By contrast, patient prognosis is a complex prediction problem, which can be driven by both intrinsic (e.g., host anti-tumor immunity and performance status) and extrinsic factors (e.g., treatment).

Pathology slide offers an avenue to quantify both tumor and host-intrinsic properties. However, its huge size and high intratumor heterogeneity have imposed significant challenges on distilling imaging patterns for survival prediction. Existing WSI feature extraction techniques can be categorized into two main paradigms: the patch-based approach [21] and the nuclei-based approach [13]. The patch-based method first divides WSIs into regular patches, then extracts patch level features using pre-trained convolutional neural network (CNN) models, and finally develops prediction models using multiple-instance learning (MIL) in combination with attention mechanisms [3,21]. One major limitation of patch-based approach is the black-box mechanism when extracting patch-wise features, which is lack of biological explanation. On the other hand, the nuclei-based approach aims to characterize the diverse cell population as well as their spatial interactions, which has a transparent interpretation. The framework typically starts with nuclei segmentation, followed by characterizing nuclei's morphology features [7,14] or constructing nuclei-level graph models to quantify intratumor heterogeneity [13]. However, graph models need to tune many hyperparameters and are computationally expensive. Thus they mainly apply to the predefined region-of-interest (ROI) on WSIs.

Here, we aim to advance the nuclei pipeline by proposing an informative embedded map based on the nuclei's types (tumor, immune, or miscellaneous) and their geographical locations, and subsequently profiling these cells' composition and architecture on WSIs to assess their implication of prognosis. In particular, we have implemented our pipeline on the TCGA lung adenocarcinoma dataset and demonstrated promising performance. This general framework has the potential to predict survival and treatment response across different cancer types.

2   Method

Figure 1 illustrates the proposed pipeline of slide embedding and its application in survival prediction, which mainly contains nuclei segmentation and classification on WSIs, embedding WSIs to smaller maps to summarize different cells' composition and architecture, texture feature extraction from embedded maps, and the machine learning model for survival prediction.

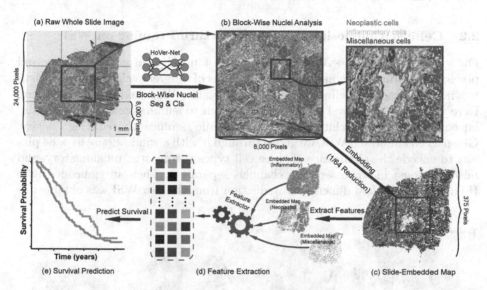

Fig. 1. The pipeline of encoding a whole slide image (WSI) into an embedded map and extracting texture features for survival prediction. (a) Splitting WSI into multiple blocks; (b) Block-wise nuclei segmentation and classification; (c) Encoding WSIs to embedded maps based on nuclei's locations and classification; (d) Extracting features from embedded maps; (e) Survival prediction based on extracted features.

2.1   Nuclei Segmentation and Classification on WSIs

We adopted a transfer learning through the HoVer-Net [9] pre-trained on the PanNuke dataset [8] for nuclei segmentation and classification. The common way to analyze WSIs is to first detect tissue regions and then apply machine learning models to the detected regions. However, this way has certain practical challenge, where most exiting tissue detection algorithms cannot generalize well on WSIs, and as such, manual hyperparameters tuning is needed to obtain reasonable results. To address this, we proposed a block-based approach to cope with large WSIs. For each given WSI, we first split it into smaller blocks (Fig. 1a) and here we defined the block size to be 8k × 8k pixels. Then we applied the HoVer-Net to these blocks separately (Fig. 1b) to segment and label cells.

After obtaining block-wise outputs, we stitch them together for the cell classification maps on WSIs. Based on the PanNuke dataset, the cells on the lung adenocarcinoma slides were classified into six categories, including neoplastic, non-neo epithelial, inflammatory, connective, dead, and non-nuclei [8]. Here, we excluded the non-nuclei category, kept the neoplastic and inflammatory, and grouped non-neo epithelial, connective and dead into the "miscellaneous" category. Then segmented nuclei are labeled into three categories, namely neoplastic (tumor), inflammatory (immune), and miscellaneous (Fig. 1b).

2.2   Cellular Composition and Architecture Profiling on WSIs

For a WSI scanned at ×20 (0.50 µm/pixel), it usually surpasses ten thousand pixels in each dimension and embodies millions of different cells. It remains challenging to extract the cellular features from gigantic WSIs and to visualize them. Here we proposed a novel embedding approach to summarize the pivotal cellular composition and architecture on WSIs, while significantly reducing the size. Given a cell annotated WSI, we glided through it with a window (size 64 × 64 pixels) to encode the abundance of three cell types (neoplastic, inflammatory, and miscellaneous) in three separate channels separately. Then an embedded map (Fig. 1c) with 64-fold dimensional reduction from the raw WSI was obtained.

(a) Raw Whole Slide Images

(b) Slide Embedded Maps

Fig. 2. Examples of raw whole slide images and their embedded maps.

We can adjust the embedding window based on the sizes of WSIs to balance between the preservation of local cellular architecture and the embedding efficiency. For instance, when large glass slides are scanned at ×40, we can further increase the window size to obtain reasonably sized embedded maps for downstream analysis. Although the embedding process compresses certain degree of local nuclei information, it offers a holistic view of the detailed global architectures of different cell population as shown in Fig. 2.

2.3   Global Cell Architecture Profiling and Prediction Model for Survival

Given the embedded maps with dimension comparable to natural images, the existing feature extraction and prediction tools developed for natural images can be directly applied to them. To quantify the intratumoral heterogeneity from these maps, we employed the texture analysis to derive the gray-level co-occurrence matrix (GLCM) features [10]. In details, the GLCM texture analysis was used to extract features on the three channels (neoplastic, inflammatory, and miscellaneous) separately (Fig. 1d). For each channel, 24 GLCM features were obtained, including: autocorrelation, joint average, cluster prominence, cluster shade, cluster tendency, contrast, correlation, difference average, difference entropy, difference variance, joint energy, joint entropy, informational measure of correlation (IMC) 1, informational measure of correlation (IMC) 2, inverse difference moment (IDM), maximal correlation coefficient (MCC), inverse difference moment normalized (IDMN), inverse difference, inverse difference normalized (IDN), inverse variance, maximum probability, sum average, sum entropy, and sum of squares [17]. These GLCM features have been widely used in radiomics studies to characterize tumor heterogeneity at the tissue level [20]. Totally, we obtained 72 GLCM features to quantify the cell heterogeneity for WSIs.

We then performed survival analysis on the extracted GLCM features using the Cox proportional hazard (CPH) model (Fig. 1e). The clinical endpoint of this study is overall survival (OS), defined as the time from the cancer diagnosis until death. Patients alive at the last follow-up are censored. Based on the quantitative risk score (i.e., the CPH model predicted hazard), we stratified the patients into low-risk and high-risk groups using an optimal cut-off derived from the quantile classification scheme [2]. The cut-off value was measured based on the training set, and then identically applied to testing set. We employed the Kaplan-Meier estimator and the two-sided logrank test to evaluate the significance of patient stratification and used Antolini's concordance index (C-index) to measure the goodness of fit for the model's discrimination power [1].

3   Experiments and Results

3.1   Dataset Description

We evaluated the proposed pipeline on the TCGA lung adenocarcinoma dataset from the TCGA data portal[1]. We retrieved the survival and phenotype information from UCSC Xena[2]. The study concentrated on WSIs scanned at ×20 with available overall survival (OS) data. After the data curation and quality control, we identified 220 patients with 344 WSIs. We split patients into training and testing cohorts using the propensity score matching (PSM) algorithm to ensure a balanced splitting, where training and testing cohorts had similar characteristics of gender, age, race, TNM stages, overall stage, and OS [16]. The splitting

---

[1] https://portal.gdc.cancer.gov/.
[2] https://tcga.xenahubs.net.

6   P. Chen et al.

ratio was 7:3, and we obtained 153 patients with 236 slides for training and 67 patients with 108 slides for testing.

3.2 Results and Discussion

We first utilized all extracted GLCM features (n = 72) to model patients' risk of death by stratifying them into the high- vs. low-risk groups and denoted it as the baseline model. As shown in Fig. 3a, we observe that the baseline model manage to stratify patients with significant difference in OS with hazard ratio (HR): 2.37 (1.51–3.71), p = 0.00012, C-index = 0.65 on training and HR: 2.01 (1.01–3.99), p = 0.039, C-index = 0.63 on testing.

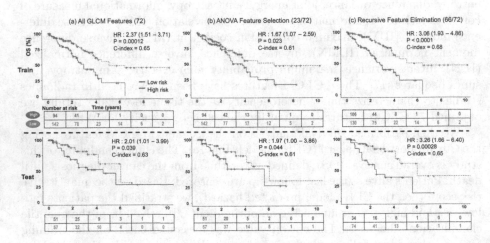

Fig. 3. Survival analysis via all extracted GLCM features and two feature selection manners. (a) Survival analysis with all 72 GLCM features; (b) Feature selection based on ANOVA analysis; (c) Feature selection via rank-based recursive feature elimination.

Next, two feature selection algorithms were evaluated to investigate whether removing redundant or less informative features while retaining those with high prognostic value would improve patient stratification. The first approach is based on analysis of variance (ANOVA), that measures the correlation of individual features to the OS in the univariate analysis. We assumed those highly informative features with p-values less than 0.05 and kept 23 out of 72 features for model development. The performance of the ANOVA selected features is shown in Fig. 3b, which has slightly inferior performance compared to the baseline model. A possible reason is that ANOVA measures the feature importance by only correlating it to binary labels (alive or dead), while ignoring the event's duration. Besides, we have removed nearly 70% features, while some of them might have complementary prognostic values. The second approach is a rank-based recursive feature elimination (RFE). This technique firstly rearranges all the features from the least to the most informative ones based on their p-values.

Then it removes one feature at a time according to its ranking orders and refits the CPH model. If the refitted model's performance doesn't drop compared to the previous model, we will remove this feature. The RFE algorithm has achieved the HR value of 3.26 on the testing cohort when eliminating 6 features, which outperforms the baseline model (Fig. 3c).

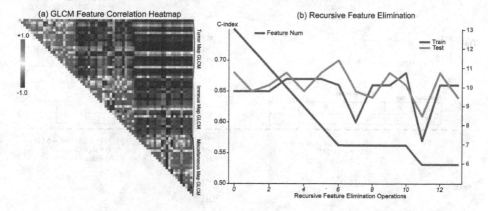

Fig. 4. GLCM feature redundancy removal analysis. (a) GLCM feature correlation matrix; (b) Survival analysis performance evolution with the recursive feature elimination.

Furthermore, GLCM features usually contain redundant information, which might lead to its performance drop during validation. To eliminate dependencies and collinearity between GLCM features, we conducted correlation analysis on GLCM features and showcased the correlation matrix in Fig. 4a. When the absolute value of the correlation coefficient between two features is higher than 0.9, we dropped the feature with a higher p-value. 13 out of 72 GLCM features were retained for survival modeling.

Figure 5a presents the survival model based on 13 GLCM features with HR: 2.79, C-index = 0.65 on training and HR: 2.54, C-index = 0.68 on testing. This suggested the benefits of removing feature redundancy on improving the survival model performance. We further applied the RFE to refine the model, which selected 7 out of 13 features and achieved the optimal performance on both training and testing cohorts, as shown in Fig. 4b. The CPH model can be defined as:

$$
\begin{aligned}
\ln \frac{h(t)}{h_0(t)} = &-0.23 * I_{Correlation} - 0.07 * N_{Autocorrelation} - 0.02 * N_{IDMN} \\
&+ 0.07 * I_{Autocorrelation} + 0.12 * I_{ClusterProminence} \\
&+ 0.14 * N_{ClusterProminence} + 0.23 * N_{InverseVariance},
\end{aligned}
\tag{1}
$$

where N and I represent neoplastic (tumor) and inflammatory (immune) embedded maps, respectively. Interestingly, all 7 selected features quantifying the heterogeneity derived from tumor and immune cell maps. Besides, the correlation of

8   P. Chen et al.

the immune map ($I_{Correlation}$) is markedly associated with improved survival, and the tumor map's inverse variance ($N_{InverseVariance}$) is mostly associated with decreased survival.

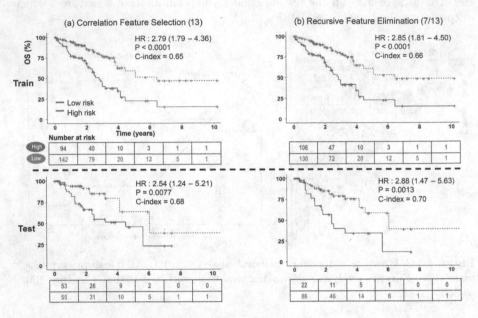

Fig. 5. Survival analysis after removing feature redundancy. (a) Survival analysis on remained features after correlation-based feature filtering; (b) Recursive feature elimination on correlation-filtered features.

There were several limitations to our study. First, the WSI nuclei segmentation and classification have high computational cost. It normally took around 2 h to complete one WSI with a dimension of $64,000 \times 48,000$ pixels with DGX A100 server (8 NVIDIA A100 GPUs). Second, nuclei segmentation and classification performance are the cornerstones of the effectiveness of the embedded maps. Most reported nuclei segmentation performances from HoVer-Net are robust and decent, with a Dice coefficient score higher than 0.8 [7,9]. While the performance of these nuclei classification expects further improvements. Third, the cohort of TCGA lung adenocarcinoma adopted for the pipeline evaluation is relatively small. The prognostic models shall be evaluated in large prospective cohorts for more rigorous assessments. Forth, the underlying biology of these pathomics features is yet to be determined. Analysis of available genomic and transcriptomic pathways of high- vs. low-risk populations may provide deeper biological insight. At last, the features we extracted from these embedded maps only scratch its surface to profile the tumor microenvironment. GLCM features extracted from the cell architectural maps only quantify three cell categories independently. In the future, we will focus on the co-localization of interested cell populations to further mine clinically relevant features from these maps.

4 Conclusion

We present a computational pipeline that can generate an embedded map to flexibly profile the local and global composition and architecture of different cell populations on WSIs. Our approach allows researchers to investigate different cell populations as well as tumor microenvironment based on these embedded maps with a reasonable size, rather than dealing with gigantic WSIs. Thus, the existing feature extraction and computer vision tools developed for natural images can directly apply to these maps. As a proof-of-concept study, we implemented the GLCM texture analysis. These extracted texture features have demonstrated promising survival prediction performance based on the analysis of TCGA lung adenocarcinoma dataset. These intriguing findings warrant future studies on applying this framework to grading and treatment response in lung and other malignancies.

References

1. Antolini, L., Boracchi, P., Biganzoli, E.: A time-dependent discrimination index for survival data. Stat. Med. **24**(24), 3927–3944 (2005)
2. Budczies, J., et al.: Cutoff finder: a comprehensive and straightforward web application enabling rapid biomarker cutoff optimization. PLoS ONE **7**(12), e51862 (2012)
3. Chen, P., Liang, Y., Shi, X., Yang, L., Gader, P.: Automatic whole slide pathology image diagnosis framework via unit stochastic selection and attention fusion. Neurocomputing **453**, 312–325 (2021)
4. Chen, P., Shi, X., Liang, Y., Li, Y., Yang, L., Gader, P.D.: Interactive thyroid whole slide image diagnostic system using deep representation. Comput. Methods Programs Biomed. **195**, 105630 (2020)
5. Diao, S., et al.: Computer-aided pathologic diagnosis of nasopharyngeal carcinoma based on deep learning. Am. J. Pathol. **190**(8), 1691–1700, e51862 (2020)
6. Diao, S., et al.: Weakly supervised framework for cancer region detection of hepatocellular carcinoma in whole-slide pathologic images based on multiscale attention convolutional neural network. Am. J. Pathol. **192**(3), 553–563 (2022)
7. El Hussein, S., et al.: Artificial intelligence strategy integrating morphologic and architectural biomarkers provides robust diagnostic accuracy for disease progression in chronic lymphocytic leukemia. J. Pathol. **256**(1), 4–14 (2022)
8. Gamper, J., Alemi Koohbanani, N., Benet, K., Khuram, A., Rajpoot, N.: PanNuke: an open pan-cancer histology dataset for nuclei instance segmentation and classification. In: Reyes-Aldasoro, C.C., Janowczyk, A., Veta, M., Bankhead, P., Sirinukunwattana, K. (eds.) ECDP 2019. LNCS, vol. 11435, pp. 11–19. Springer, Cham (2019). https://doi.org/10.1007/978-3-030-23937-4_2
9. Graham, S., et al.: HoVer-Net: simultaneous segmentation and classification of nuclei in multi-tissue histology images. Med. Image Anal. **58**, 101563 (2019)
10. Haralick, R.M., Shanmugam, K., Dinstein, I.H.: Textural features for image classification. IEEE Trans. Syst. Man Cybern. **6**, 610–621 (1973)
11. Hekler, A., et al.: Pathologist-level classification of histopathological melanoma images with deep neural networks. Eur. J. Cancer **115**, 79–83 (2019)

12. Li, R., Yao, J., Zhu, X., Li, Y., Huang, J.: Graph CNN for survival analysis on whole slide pathological images. In: Frangi, A.F., Schnabel, J.A., Davatzikos, C., Alberola-López, C., Fichtinger, G. (eds.) MICCAI 2018. LNCS, vol. 11071, pp. 174–182. Springer, Cham (2018). https://doi.org/10.1007/978-3-030-00934-2_20

13. Lu, C., et al.: Feature-driven local cell graph (FLocK): new computational pathology-based descriptors for prognosis of lung cancer and HPV status of oropharyngeal cancers. Med. Image Anal. **68**, 101903 (2021)

14. Lu, C., Lewis, J.S., Dupont, W.D., Plummer, W.D., Janowczyk, A., Madabhushi, A.: An oral cavity squamous cell carcinoma quantitative histomorphometric-based image classifier of nuclear morphology can risk stratify patients for disease-specific survival. Mod. Pathol. **30**(12), 1655–1665 (2017)

15. Lu, M.Y., et al.: AI-based pathology predicts origins for cancers of unknown primary. Nature **594**(7861), 106–110 (2021)

16. Rosenbaum, P.R., Rubin, D.B.: The central role of the propensity score in observational studies for causal effects. Biometrika **70**(1), 41–55 (1983)

17. Van Griethuysen, J.J., et al.: Computational radiomics system to decode the radiographic phenotype. Can. Res. **77**(21), e104–e107 (2017)

18. Viswanathan, V.S., Toro, P., Corredor, G., Mukhopadhyay, S., Madabhushi, A.: The state of the art for artificial intelligence in lung digital pathology. J. Pathol. **257**, 413–429 (2022)

19. Wei, J.W., Tafe, L.J., Linnik, Y.A., Vaickus, L.J., Tomita, N., Hassanpour, S.: Pathologist-level classification of histologic patterns on resected lung adenocarcinoma slides with deep neural networks. Sci. Rep. **9**(1), 1–8 (2019)

20. Wu, J., Mayer, A.T., Li, R.: Integrated imaging and molecular analysis to decipher tumor microenvironment in the era of immunotherapy. In: Seminars in Cancer Biology. Elsevier (2020)

21. Yao, J., Zhu, X., Jonnagaddala, J., Hawkins, N., Huang, J.: Whole slide images based cancer survival prediction using attention guided deep multiple instance learning networks. Med. Image Anal. **65**, 101789 (2020)

22. Zhang, Z., et al.: Pathologist-level interpretable whole-slide cancer diagnosis with deep learning. Nat. Mach. Intell. **1**(5), 236–245 (2019)

Is More Always Better? Effects of Patch Sampling in Distinguishing Chronic Lymphocytic Leukemia from Transformation to Diffuse Large B-Cell Lymphoma

Rukhmini Bandyopadhyay[1], Pingjun Chen[1], Siba El Hussein[2],
Frank R. Rojas[3], Kingsley Ebare[4], Ignacio I. Wistuba[3], Luisa M. Solis Soto[3],
L. Jeffrey Medeiros[5], Jianjun Zhang[6,7], Joseph D. Khoury[8], and Jia Wu[1,6(✉)]

[1] Department of Imaging Physics, The University of Texas MD Anderson Cancer
Center, Houston, TX 77030, USA
jwu11@mdanderson.org
[2] Department of Pathology, University of Rochester Medical Center,
Rochester, NY 14642, USA
[3] Department of Translational Molecular Pathology,
The University of Texas MD Anderson Cancer Center, Houston, TX 77030, USA
[4] Department of Pathology, The University of Texas MD Anderson Cancer Center,
Houston, TX 77030, USA
[5] Department of Hematopathology, The University of Texas MD Anderson Cancer
Center, Houston, TX 77030, USA
[6] Department of Thoracic/Head and Neck Medical Oncology,
The University of Texas MD Anderson Cancer Center, Houston, TX 77030, USA
[7] Department of Genomic Medicine, The University of Texas MD Anderson Cancer
Center, Houston, TX 77030, USA
[8] Department of Pathology and Microbiology, University of Nebraska Medical
Center, Omaha, NE 68198, USA

Abstract. Distinguishing chronic lymphocytic leukemia (CLL), accelerated phase of CLL (aCLL), and diffuse large B-cell lymphoma transformation of CLL (Richter transformation; RT) has important clinical
implications that greatly influence patient management. However, distinguishing between these disease phases on histologic grounds may be
challenging in routine practice due to the presence of similar structures
and homogeneous intensity, among others. In this work, we propose a
whole slide image (WSI) level computational framework based on the
integration of deep transfer learning, patch level random sampling, and
machine learning modeling to distinguish CLL from aCLL and RT. The
motivation behind the proposed random sampling-based classification
is to address a fundamental question in WSI analysis: is it true that
more data is always better? To answer this question, we apply this
framework on a pilot cohort of 56 patients (total 95 WSIs). Interestingly, we observe that the tested machine learning models demonstrate a

R. Bandyopadhyay and P. Chen—Equal Contribution.

© The Author(s), under exclusive license to Springer Nature Switzerland AG 2022
W. Qin et al. (Eds.): CMMCA 2022, LNCS 13574, pp. 11–20, 2022.
https://doi.org/10.1007/978-3-031-17266-3_2

robust performance with just 1% randomly sampled patches from WSIs, on par with the model built on the entire WSI data. Among all three tested machine learning algorithms, multi-instance learning (MIL) has achieved the best prediction, outperforming SVM and XGBoost models. Taken together, our pilot study shows that machine learning models can potentially achieve a reasonable performance with a substantially lower amount of data from WSIs. This observation will shed light on shaping future WSI analysis, where we may reduce the computational burden by using fewer numbers of patches rather than all the data in WSIs, thereby improving computational efficiency. However, these results need to be validated and cautiously interpreted, where the findings may be fundamentally driven by the homogeneous appearance of CLL in pathology slides. It remains unclear if this finding will hold up when testing is performed on more heterogeneous cancer types.

Keywords: Chronic Lymphocytic Leukemia (CLL) · Accelerated CLL · Richter Transformation (RT) · Random sampling · Patch level analysis

1 Introduction

Chronic lymphocytic leukemia (CLL) is the most common leukemia in western countries and accounts for one-third of new leukemia cases every year [16]. CLL is usually a slowly progressive disease but it can transform into a more clinically aggressive diffuse large B-cell lymphoma (DLBCL) in 10% of patients. This transformation is also known as Richter transformation (RT) [1]. An intermediate stage of disease progression known as accelerated phase CLL (aCLL) has overlapping characteristics of both CLL and RT, which makes its diagnosis challenging [5]. Accurate diagnosis/classification of CLL, aCLL and RT is clinically important as this decision helps to determine the necessity of escalating chemo-immunotherapy based on the disease progression.

In recent years, machine learning-based algorithms, especially deep learning, have shown the potential for analyzing digital pathology slides. The applications range from cell segmentation and subtyping to the detection of disease progression [2,10,14,23,27]. These algorithms perform disease diagnosis via either handcrafted pathomic features [8], or deep learning distilled features [3,21], or even synergism of hand-crafted and deep features [13]. There are a few pilot studies that have applied machine learning algorithms to the analysis of disease progression of CLL [4,5,12,13,18]. However, these studies hinge on pathologist annotated region-of-interests (ROI). In contrast, few studies use whole slide images (WSI) to classify CLL progression [25].

The patch-based approach is widely adopted to analyze WSIs [7,17,19], which first splits WSIs into multiple patches, next performs inference on individual patches, and then aggregates patch predictions for the WSI prediction via multi-instance learning (MIL) [6,11,20,26]. For a gigapixel WSI, the number of split patches can easily surpass 10,000, which causes enormous computational burden and delay the diagnosis speed. Currently, there are limited studies on the effects of patch sampling on WSI diagnosis.

In this study, we propose a novel computational pipeline for automated classification of CLL, aCLL, and RT, and systematically investigate the effects of patch sampling on WSI diagnosis. The main contributions are:

1. For WSI analysis, we aim to address a more central question: is it true that more data is better? We seek to determine the minimum set of data needed to distinguish CLL disease progression. Here, we apply patch level sampling from WSIs at different percentages and investigate its relationship with the model performance of classifying CLL, aCLL and RT.
2. From the application perspective, we have applied automated WSI level analysis for the classification of disease progression in terms of CLL, aCLL and RT. By contrast, most previous studies relied on the selectively annotated ROIs.

2 Proposed Methodology

In this section, we describe the methodology in detail and the overview of the proposed pipeline is depicted in Fig. 1. It comprises of three modules, (a) WSI preprocessing (b) Patch-level random sampling, and (c) WSI Classification.

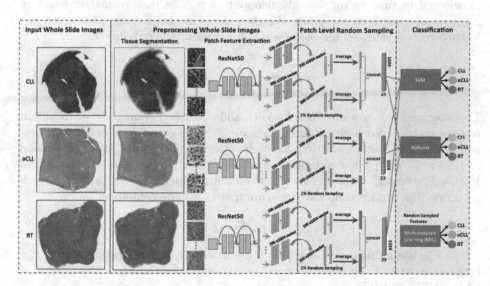

Fig. 1. Overview of the proposed pipeline: WSI patch features are extracted using ResNet50; patch level random sampling is performed to reduce the number of patches used for the WSI feature generation; Machine learning algorithms are trained and evaluated for the classification of CLL, aCLL, and RT.

2.1 WSI Preprocessing

The first step of our proposed method involves preprocessing of WSIs using clustering-constrained-attention MIL (CLAM) [20]. For each WSI in the three categories of CLL, aCLL and RT, an automated segmentation of the tissue region is applied based on thresholding, and additionally the small gaps and holes are filled by morphological closing operators. The foreground regions on the WSIs are then stored for downstream processing. After segmentation, the patches of size 1024×1024 are extracted from the foreground region. The patches along with their co-ordinates are stored for further processing. After the patch extraction, a pretrained deep CNN model ResNet50 is applied to convert each 1024×1024 patch into a 1024-dimensional feature vector [15]. The obtained feature vectors are subsequently used for our proposed patch-level random sampling approach.

2.2 Patch-Level Random Sampling

Let, k be the number of patches extracted from each individual WSI of CLL, aCLL and RT and F_k be the feature matrix of size $1024 \times k$ obtained from ResNet50. The number of patches k extracted from each slide for each category (CLL, aCLL, and RT) varies from thousands to tens of thousands (resection slide scanned at ×20 magnification). Let, $n\%$ be the sampling ratio which is used to randomly sample patch features from F_k and results in the reduced number of selected feature vector $k \times n\%$ denoted as p. The feature matrix becomes $F_{1024 \times p}$ after random sampling. We then perform the average pooling on the feature vectors sampled for each slide and represent the outcome as,

$$F_{avg} = \frac{1}{p} \sum_{i=1}^{p} F_{1024 \times i}. \tag{1}$$

The averaged features obtained for each slide are then concatenated to obtain feature matrices for each category. Let, F_{CLL}, F_{aCLL} and F_{RT} be the concatenated averaged feature matrices for each category where F_{CLL} has a dimension of $1024 \times C$, F_{aCLL} has a dimension of $1024 \times A$ and F_{RT} has a dimension of $1024 \times R$ where C, A and R denote the total number of WSIs present in each category. The final concatenated feature matrix is represented as,

$$F_{all} = [F_{CLL}\ F_{aCLL}\ F_{RT}], \tag{2}$$

where F_{all} has a dimension of $1024 \times (C + A + R)$. Now, the feature matrix F_{all} is fed as input to train different machine learning algorithms for classification of CLL, aCLL, and RT.

2.3 WSI Classification

The proposed patch level random sampling approach is evaluated by classification into three categories typical CLL, aCLL and RT using three different machine learning approaches, including SVM [24], XGBoost [9], and MIL [20]. For SVM and XGBoost, we train the classifiers on the random sampled and average pooling feature matrix F_{all} from the whole feature set, and for MIL, we use random sampled patch level features without average pooling as input instances. In addition, we repeat the random sampling and machine learning modeling multiple times to evaluate its robustness.

3   Experimental Analysis

3.1   Dataset Description

For this study, a group of 56 patients with 95 specimens are included. Among these 95 slides, 43 slides are from 23 CLL patients, 23 slides are from 16 aCLL patients and 29 slides are from 17 RT patients. Routine hematoxylin and eosin (H&E) slides are scanned using Aperio AT2 scanners at an optical resolution of ×20 (0.50 μm/pixel). Scanning is performed in three batches within the same day, using the same scanner and settings.

3.2   Results and Discussion

First, we aim to check the effects of random sampling on patient-level stratification. Given the average pooled patch features for individual patients, we embed the whole population to the latent feature space with t-SNE [22], as shown in Fig. 2(a). Without loss of generality, we randomly sample 1% patches from each patient's WSI to embed them in the latent feature space, and we repeat this procedure 100 times which generates a population for three disease categories. Next, we combine the mean of the whole population with 1% sampled patches repeated 100 times to observe their alignment. The combined t-SNE plot in Fig. 2(b) signifies that mean of the complete feature points are well aligned with the 1% random sampled feature population of each category.

We perform a 3-fold cross-validation to train and test three machine learning models including CLAM, SVM and XGBoost. During the training and testing cohort splitting, we make sure that the slides from the same patient are only present in one cohort to prevent patient level information leakage. The 3-fold cross-validation without random sampling for SVM and XGBoost is used as a baseline for comparison. The first two splits contain 63 training and 32 testing slides and the last split contains 64 training and 31 testing slides. In MIL using CLAM, the first split contains 61 training, 9 validation and 25 testing slides. The second split contains 71 training, 9 validation and 15 testing slides and the third split contains 62 training, 11 validation and 22 testing slides. The performance is reported on the testing slides for SVM, XGBoost, and CLAM.

In Fig. 3 we compare the performance of SVM, XGBoost and CLAM for five different patch sampling. We gradually increase the random patch selection from 5 patches, 10 patches, 0.1%, 0.5%, and 1%, and then repeat for 10 times in the 3-fold cross-validation. We observe that the performance of individual models

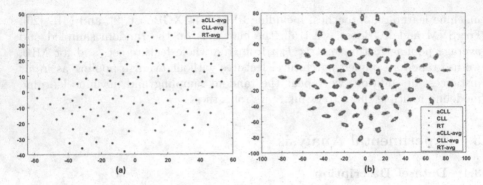

Fig. 2. Comparison of t-SNE plot: (a) Average t-SNE plot without random sampling; (b) t-SNE plot with 1% random sampling repeated 100 times, combined with average t-SNE plot.

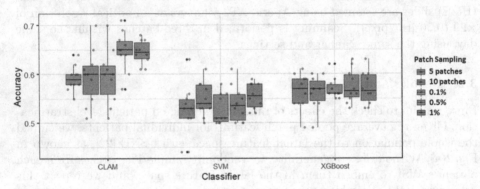

Fig. 3. Performance comparison of CLAM, SVM, and XGBoost on different patch sampling strategies.

improves when feeding more patches. Interestingly, when 1% patches are sampled, the machine learning models start to saturate and reach the performance of models trained with all WSI patches. Besides, the MIL approach (CLAM) has significantly outperformed the classical SVM and XGBoost, possibly due to its superior way of handling the patches in a weakly supervised manner.

In Fig. 4, we demonstrate the classification accuracy of SVM for CLL, aCLL and RT using the confusion matrix for each fold of the 3-fold cross-validation. In particular, we show the classification accuracy with 1% random sampling (first row in Fig. 4) and without random sampling (second row in Fig. 4) which gives the overall average accuracy of 0.57 and 0.58 respectively. Similarly, overall average accuracy of XGBoost algorithm with 1% random sampling (first row in Fig. 5) and without random sampling (second row in Fig. 5) are 0.58 and 0.59, respectively. Also, the overall average accuracy of CLAM algorithm with 1% random sampling (first row in Fig. 6) and without random sampling (second row in Fig. 6) are 0.65 and 0.62, respectively. The observed equivalent classification

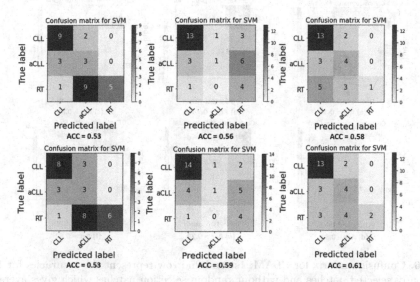

Fig. 4. Confusion matrix for SVM: 1st and 2nd row represent the accuracies for 1% randomly selected patches and without random selection patches which gives average accuracy of 0.56 and 0.58 respectively; column represents the confusion matrix for the 3-fold cross-validation.

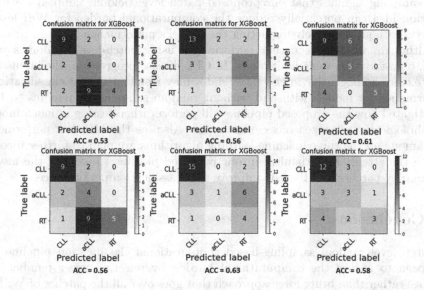

Fig. 5. Confusion matrix for XGBoost: 1st and 2nd row represent the accuracies for 1% randomly selected patches and without random selection patches, which gives average accuracy of 0.57 and 0.59 respectively; column represents the confusion matrix for the 3-fold cross-validation.

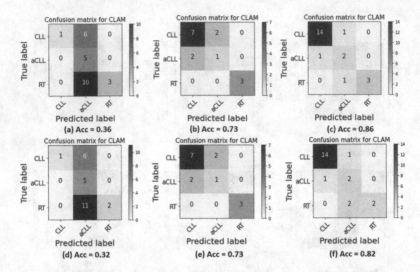

Fig. 6. Confusion matrix for CLAM: 1st and 2nd row represent the accuracies for 1% randomly selected patches and without random selection patches which gives average accuracy of 0.65 and 0.62 respectively; column represents the confusion matrix for the 3-fold cross-validation.

performance between individual machine learning models with and without random sampling signifies that our proposed patch level random sampled feature selection idea can potentially reduce the computational burden for WSI level analysis without negatively impacting the model's performance.

Although, we achieve decent performance using patch-level random sampling, our study has some limitations. In CLL disease progression, the intensity and texture homogeneity in the H&E slides may result in similar classification performance for models with and without random sampling. It remains to be investigated how the proposed pipeline will perform when testing in much more morphologically heterogeneous cancer types. Also, as the MIL has performed best among all the three machine learning algorithms, we plan to further incorporate the attention mechanism in our proposed pipeline to identify the most informative patches and thereby improve the classification performance.

4   Conclusion

A patch level random sampling-based computational classification pipeline is proposed to reduce the computational burden by selecting fewer number of patches rather than brute-force approach that goes over all the patches of WSIs. We have demonstrated the power of this framework to diagnose progression and transformation of CLL into aCLL and RT. Among the three machine learning algorithms, MIL has the optimal performance. It performs robustly with just 1% randomly sampled patches from the WSIs, similar to the model trained with all available data. Taken together, this approach paves a way for us to efficiently handle the WSI level analysis in blood cancer and other cancer types.

References

1. Agbay, R.L.M.C., Jain, N., Loghavi, S., Medeiros, L.J., Khoury, J.D.: Histologic transformation of chronic lymphocytic leukemia/small lymphocytic lymphoma. Am. J. Hematol. **91**(10), 1036–1043 (2016)
2. Ailia, M.J., Thakur, N., Abdul-Ghafar, J., Jung, C.K., Yim, K., Chong, Y.: Current trend of artificial intelligence patents in digital pathology: a systematic evaluation of the patent landscape. Cancers **14**(10), 2400 (2022)
3. Braman, N., Gordon, J.W.H., Goossens, E.T., Willis, C., Stumpe, M.C., Venkataraman, J.: Deep orthogonal fusion: multimodal prognostic biomarker discovery integrating radiology, pathology, genomic, and clinical data. In: de Bruijne, M., et al. (eds.) MICCAI 2021. LNCS, vol. 12905, pp. 667–677. Springer, Cham (2021). https://doi.org/10.1007/978-3-030-87240-3_64
4. Chen, P., Aminu, M., El Hussein, S., Khoury, J.D., Wu, J.: Hierarchical phenotyping and graph modeling of spatial architecture in lymphoid neoplasms. In: de Bruijne, M., et al. (eds.) MICCAI 2021. LNCS, vol. 12908, pp. 164–174. Springer, Cham (2021). https://doi.org/10.1007/978-3-030-87237-3_16
5. Chen, P., et al.: Chronic lymphocytic leukemia progression diagnosis with intrinsic cellular patterns via unsupervised clustering. Cancers **14**(10), 2398 (2022)
6. Chen, P., Liang, Y., Shi, X., Yang, L., Gader, P.: Automatic whole slide pathology image diagnosis framework via unit stochastic selection and attention fusion. Neurocomputing **453**, 312–325 (2021)
7. Chen, P., Xie, Y., Hai, S., Yang, L.: Automatic pathology diagnosis on large slide image using patch aggregation. In: Laboratory Investigation, vol. 98, pp. 586–586. Nature Publishing Group, New York (2018)
8. Chen, R.J., et al.: Pathomic fusion: an integrated framework for fusing histopathology and genomic features for cancer diagnosis and prognosis. IEEE Trans. Med. Imaging **41**, 757–770 (2020)
9. Chen, T., Guestrin, C.: XGBoost: a scalable tree boosting system. In: Proceedings of SIGKDD, pp. 785–794 (2016)
10. Diao, S., et al.: Computer-aided pathologic diagnosis of nasopharyngeal carcinoma based on deep learning. Am. J. Pathol. **190**(8), 1691–1700 (2020)
11. Diao, S., et al.: Weakly supervised framework for cancer region detection of hepatocellular carcinoma in whole-slide pathologic images based on multiscale attention convolutional neural network. Am. J. Pathol. **192**(3), 553–563 (2022)
12. El Hussein, S., Chen, P., Medeiros, L.J., Hazle, J.D., Wu, J., Khoury, J.D.: Artificial intelligence-assisted mapping of proliferation centers allows the distinction of accelerated phase from large cell transformation in chronic lymphocytic leukemia. Mod. Pathol. **35**, 1–5 (2022)
13. El Hussein, S., et al.: Artificial intelligence strategy integrating morphologic and architectural biomarkers provides robust diagnostic accuracy for disease progression in chronic lymphocytic leukemia. J. Pathol. **256**(1), 4–14 (2022)
14. Graham, S., et al.: HoVer-Net: simultaneous segmentation and classification of nuclei in multi-tissue histology images. Med. Image Anal. **58**, 101563 (2019)
15. He, K., Zhang, X., Ren, S., Sun, J.: Deep residual learning for image recognition. In: Proceedings of the IEEE Conference on Computer Vision and Pattern Recognition, pp. 770–778 (2016)

16. Hortal, A.M., et al.: Overexpression of wild type RRAS2, without oncogenic muta-
    tions, drives chronic lymphocytic leukemia. Mol. Cancer **21**(1), 1–24 (2022)
17. Hou, L., Samaras, D., Kurc, T.M., Gao, Y., Davis, J.E., Saltz, J.H.: Patch-based
    convolutional neural network for whole slide tissue image classification. In: Pro-
    ceedings of the IEEE Conference on Computer Vision and Pattern Recognition,
    pp. 2424–2433 (2016)
18. Li, D.: A deep learning diagnostic platform for diffuse large B-cell lymphoma with
    high accuracy across multiple hospitals. Nat. Commun. **11**(1), 1–9 (2020)
19. Li, Y., Chen, P., Li, Z., Su, H., Yang, L., Zhong, D.: Rule-based automatic diagnosis
    of thyroid nodules from intraoperative frozen sections using deep learning. Artif.
    Intell. Med. **108**, 101918 (2020)
20. Lu, M.Y., Williamson, D.F., Chen, T.Y., Chen, R.J., Barbieri, M., Mahmood,
    F.: Data-efficient and weakly supervised computational pathology on whole-slide
    images. Nat. Biomed. Eng. **5**(6), 555–570 (2021)
21. Lv, Z., Lin, Y., Yan, R., Yang, Z., Wang, Y., Zhang, F.: PG-TFNET: transformer-
    based fusion network integrating pathological images and genomic data for cancer
    survival analysis. In: 2021 IEEE BIBM, pp. 491–496. IEEE (2021)
22. Van der Maaten, L., Hinton, G.: Visualizing data using t-SNE. J. Mach. Learn.
    Res. **9**(11), 2579–2605 (2008)
23. Mahmood, F., et al.: Deep adversarial training for multi-organ nuclei segmentation
    in histopathology images. IEEE Trans. Med. Imaging **39**(11), 3257–3267 (2019)
24. Noble, W.S.: What is a support vector machine? Nat. Biotechnol. **24**(12), 1565–
    1567 (2006)
25. Steinbuss, G., et al.: Deep learning for the classification of non-hodgkin lymphoma
    on histopathological images. Cancers **13**(10), 2419 (2021)
26. Yao, J., Zhu, X., Jonnagaddala, J., Hawkins, N., Huang, J.: Whole slide images
    based cancer survival prediction using attention guided deep multiple instance
    learning networks. Med. Image Anal. **65**, 101789 (2020)
27. Zhang, Z., et al.: Pathologist-level interpretable whole-slide cancer diagnosis with
    deep learning. Nat. Mach. Intell. **1**(5), 236–245 (2019)

Repeatability of Radiomic Features Against Simulated Scanning Position Stochasticity Across Imaging Modalities and Cancer Subtypes: A Retrospective Multi-institutional Study on Head-and-Neck Cases

Jiang Zhang[1], Saikit Lam[1], Xinzhi Teng[1], Yuanpeng Zhang[2], Zongrui Ma[1], Francis Lee[3], Kwok-hung Au[3], Wai Yi Yip[3], Tien Yee Amy Chang[4], Wing Chi Lawrence Chan[1], Victor Lee[5], Q. Jackie Wu[6], and Jing Cai[1](✉)

[1] Department of Health Technology and Informatics,
The Hong Kong Polytechnic University, Hung Hom, Kowloon, Hong Kong SAR
`jing.cai@polyu.edu.hk`
[2] Department of Medical Informatics, Nantong University, Nantong, Jiangsu, China
[3] Department of Clinical Oncology, Queen Elizabeth Hospital, King's Park,
Kowloon, Hong Kong SAR
[4] Department of Clinical Oncology, Hong Kong Sanatorium & Hospital,
2 Village Road, Happy Valley, Hong Kong SAR
[5] Department of Clinical Oncology,
The University of Hong Kong, Pokfulam, Hong Kong SAR
[6] Department of Radiation Oncology, Duke University Medical Center,
Durham, NC, USA

Abstract. We attempted to investigate the Radiomic feature (RF) repeatability and its agreements across imaging modalities and head-and-neck cancer (HNC) subtypes via image perturbations. Contrast-enhanced computed tomography (CECT), CET1-weight, T2-weight magnetic resonance images of 231 nasopharyngeal carcinoma (NPC) patients, and CECT images of 399 oropharyngeal carcinoma (OPC) patients were retrospectively analyzed. Randomized translation and rotation were implemented to the images for mimicking scanning position stochasticity. 1288 RFs from unfiltered, Laplacian-of-Gaussian-filtered (LoG), and wavelet-filtered images were subsequently computed per perturbed image. The intra-class correlation coefficient (ICC) was calculated to assess RF repeatability. The mean absolute difference (MAD) of the ICC and the binarized repeatability consistency between image sets were adopted to evaluate its agreements across imaging modalities and HNC subtypes. Bias from feature collinearity was also investigated. All the shape RFs and the majority of RFs from unfiltered ($\geq$83.5%) and LoG-filtered ($\geq$93%) images showed high repeatability (ICC $\geq$ 0.9) in all studied datasets, whereas more than 50% of the wavelet-filtered RFs had low repeatability (ICC < 0.9). RF repeatability agreements between imaging modalities within the NPC cohort were outstanding (MAD < 0.05,

W. Qin et al. (Eds.): CMMCA 2022, LNCS 13574, pp. 21–34, 2022.
https://doi.org/10.1007/978-3-031-17266-3_3

consistency > 0.9) and slightly higher between the NPC and OPC cohort (MAD = 0.06, consistency = 0.89). Minimum bias from feature collinearity was observed. We urge caution when handling wavelet-filtered RFs and advise taking initiatives to exclude underperforming RFs during feature pre-selection for robust model construction.

Keywords: Radomics · Head and neck cancer · Repeatability

1   Introduction

Radiomics involves computerized extraction and analysis of a myriad of quantitative radiomic features (RFs) with high throughput from medical images, such as computed tomographic (CT) and magnetic resonance (MR) scans, for divulging cancer biologic and genetic traits [8]. It has offered enlightening insights into cancer diagnosis [4], prognostication [20], and treatment response prediction [18]. Nevertheless, the clinical applicability of these radiomic models has largely been impeded by the lack of studies assessing the RF robustness in their models [2,9]. As highlighted in several excellent review articles, repeatability and reproducibility of RFs are crucial for reaching reliable and consistent conclusions between studies [12,13,16]. In particular, high repeatability, referring to RFs that remain stable when imaged multiple times if the conditions keep unchanged [14], is the first and foremost criterion towards clinical utility. Further, identifying high or low repeatable RFs that are generalizable across different cancer subtypes will provide the radiomics community with direct perceptivity for selecting reliable radiomic features and building robust predictive models for implementing precision medicine.

Efforts with an attempt to bridge this important gap in knowledge mainly focused on test-retest patient images [6,10,21]. They have underlined the pronounced impacts of scanning position variations on RFs repeatability. Notwithstanding, there are noteworthy shortcomings in their studies. First, the impact of segmentation variations on RFs repeatability is often inherent in a test-retest study, where segmentations of region-of-interest are separately delineated on test and retest images, which hinders direct interpretations of the influences on RFs repeatability caused purely by positional discrepancies. Secondly, the prolonged time-interval between test and retest images, in the case of 2-week apart, might disregard the implications of intra-tumoral microbiologic changes during that period of time, which itself might lead to dramatic disparity in RFs between the two scans. Thirdly, the limited sample size owing to the need for recruiting consented patients renders their conclusions less statistically convincible.

To address these limitations, we attempted to deploy our in-house developed image perturbation framework, taking reference from a previous work by Zwanenburg et al. [21], to mimic a vast amount of scanning position stochasticity via large patient cohorts of nasopharyngeal carcinoma (NPC) and oropharyngeal carcinoma (OPC). To our best knowledge, the RF repeatability against scanning position variations in head and neck cancer (HNC) is yet to be explored, and there are no relevant publications with multiple imaging modalities. The main objectives of this study were (i) to ascertain the repeatability of RFs against scanning position stochasticity via image perturbations in both cohorts and (ii) to examine their generalizability across CT and MR imaging modalities among NPC patients. Meanwhile, we sought (iii) to assess their generalizability among HNC subtypes via a publicly available OPC dataset.

2 Methods and Materials

Figure 1 illustrates the overall study workflow. Two HNC cohorts were enrolled in this study: an internal NPC cohort and a publicly available OPC cohort. The NPC cohort consists of three image modalities, which are contrast-enhanced CT (CECT), contrast-enhanced T1 weighted (CET1-w) MR, and T2 weighted (T2-w) MR. Only CECT images were studied for the OPC cohort. We evaluated the RF repeatability generalizability across image modalities and cancer subtypes by multiple comparisons.

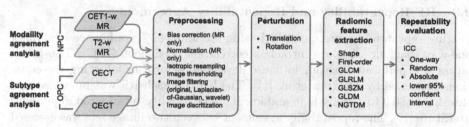

Abbreviations: NPC, Nasopharyngeal Carcinoma; OPC, Oropharyngeal Carcinoma; CET1-w MR, contrast enhanced T1-weighted magenatic resonance imaging; T2-w MR, T2-weighted megnatic resonance imaging; CECT: contrast enhanced computed tomography; GLCM, gray level cooccurrence matrix; GLRLM, gray level run length matrix; GLSZM, gray level size zone matrix; GLDM, hray level dependence matrix; NGTDM, neighboring gray tone difference matrix; ICC, intra-class correlation coefficient

Fig. 1. Overall study workflow.

2.1 Patient Cohorts

A total of 250 biopsy-proven (Stage I-IVB) NPC patients who received cancer treatment at a local hospital between 2012 and 2016 were retrospectively screened, and 231 patients that had same-institution MR images and eligible target contours were enrolled in the study. CECT images of 492 (Stage I-IV) OPC patients between 2005 and 2012 were downloaded online, and 399 patients who have eligible target contours were enrolled in this study. In the internal NPC cohort, the primary gross-tumor-volumes of NPC were manually delineated on axial CT slices co-registered with MR images by oncologists specialized in head-and-neck cancer with accreditations. In the external OPC cohort, expert radiation oncologists manually segmented primary disease gross volumes.

2.2 Image Preprocessing, Perturbation, and Feature Extraction

All the calculations in image preprocessing, perturbation, and feature extraction were performed by our in-house developed Python-based (3.7.3) pipeline using the SimpleITK (1.2.4) [1] and PyRadiomics (2.2.0) package [7]. All the image processing parameters were listed in Table 2. Image perturbations were applied to each pair of the preprocessed original-resolution image and ROI mask during isotropic resampling. Two perturbation modes, rotation ($\theta \in [-20°, 20°]$, step size = 5, around central z-axis) and translation ($\mu \in [0.00, 0.80]$, step size = 0.2, along all three dimensions), were implemented following the procedures proposed by Zwanenburg et al. [21] to mimic variations in scanning setup positions during image acquisition. In this study, 40 perturbation parameter sets (θ and μ) were randomly chosen without replacement from the 1125 possible combinations. Detailed feature extraction parameters were listed in Table 2. Feature computation was performed on the perturbed images using PyRadiomics. A total of 1288 RFs (14 shape features, 91 from the unfiltered image, and 91×13 from filtered images) was computed per perturbed image.

2.3 RF Repeatability and Repeatability Agreement

Feature repeatability was quantified using the lower 95% confidence interval of one-way, random, absolute intraclass correlation coefficient (ICC). The calculation was performed by our in-house developed algorithm following the equations presented by McGraw et al. [11]. The ICC for each RF was binarized to a threshold of 0.9 to classify high and low RF repeatability, as adopted in previous literature [17]. The repeatability agreement between two image sets was assessed using two metrics. The mean absolute difference (MAD) of the ICC was computed between the two compared datasets for each RF category, irrespective of the chosen ICC threshold. We also evaluated the RF repeatability consistency between image sets. It is quantified as the ratio of the commonly-high-/-low repeatable RFs binarized by the specified ICC threshold of 0.9.

2.4 Feature Collinearity

We analyzed the bias of our results by evaluating feature collinearity through two sub-analyses. First, we analyzed whether the inter-feature correlation affects the skewness of RF repeatability. We followed the analysis procedure proposed by Fiset et al. [5] and compared the feature repeatability distributions between all the extracted features and the independent features selected by KMeans clustering. Quantitatively, we compared the ratios of low-repeatable features (ICC < 0.9) between all the extracted features and the independent features for each image set. Second, how ROI volume dependency affects repeatability was also investigated. ROI volume is one highly repeatable feature by definition and a common prognostic factor for many disease types [3]. For each feature category and image set, the high-repeatable portion of volume-independent features was compared with the ratio relative to all the extracted features. The squared value of the Pearson correlation coefficient was used to quantify the volume correlation, and a threshold of 0.6 was chosen to determine whether an RF correlates with volume.

3 Results

3.1 RF Repeatability in Both NPC and OPC Cohorts

All the shape RFs and most unfiltered RFs (NPC:≥95.6%, OPC:83.5%) and LoG-filtered RFs (NPC:≥93.0%, OPC:93.6%) was highly repeatable against the studied positional variations, which is also visualized as the dominating blue-shaded regions in Fig. 2. However, more than half of the wavelet RFs in all the analyzed image sets had low repeatability, as shown in Table 3. Within the wavelet-filtered categories, we observed that applying high-pass wavelet filters on more dimensions or on the slice direction (from LLL to HLL/LHL to HHL/**H) caused a significant increase in low-repeatable RFs (Table 3, from 3.3–13.2% to 31.3–41.2% to 69.7–80.0%) and visualized by the increasing fractions of green-shaded regions in Fig. 2.

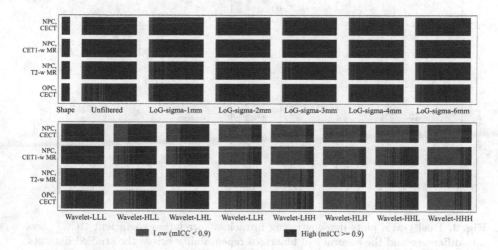

Fig. 2. Visualization of category-based feature (RF) repeatability, binarized according to a threshold of ≥ 0.9 for the median intra-class correlation coefficient (ICC). The green vertical lines represent low repeatability (ICC < 0.9) and the blue ones represent high repeatability (ICC ≥ 0.9). Within each category, features are sorted based on the ICCs of NPC CECT images and aligned at the same horizontal positions for all the image sets. (Color figure online)

3.2   Agreement of RF Repeatability Across Imaging Modalities and HNC Subtypes

For all the extracted RFs, high repeatability agreements were observed between any pair of the studied NPC image sets (MAD < 0.05, consistency > 0.9). As shown in Fig. 3a–c, shape, unfiltered, and LoG-filtered RFs expressed the highest repeatability agreements in terms of MAD (<0.02) and consistency (>0.92). Wavelet-LLL/-HLL/-LHL showed the intermediate agreement with small MAD (<0.03) but lower consistency (0.83–0.98). The remaining wavelet-filtered RFs demonstrated the lowest repeatability agreement in terms of both MAD (0.04–0.14) and consistency (0.70–0.98). The color agreements in Fig. 2 visualized such repeatability agreements. Of note, 28.5% of all the extracted RFs (367/1288) with low repeatability were commonly found across all the imaging modalities within the NPC cohort (Table 3).

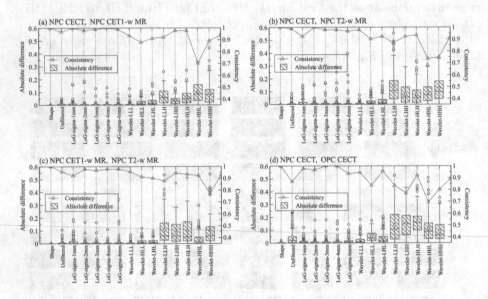

Fig. 3. Dual y-axis plots demonstrating intraclass correlation coefficient (ICC) absolute difference and the accuracy of binarized repeatability across the studied datasets. Distributions of ICC absolute difference across imaging modalities of NPC patients (a–c) and between NPC and OPC CECT images (d) are represented as blue boxes, and the repeatability accuracies using the threshold of 0.9 are drawn as green curves with triangle points. (Color figure online)

RF repeatabilities were slightly lower between CECTs of the NPC and OPC cohort (MAD = 0.06, consistency = 0.89) for all the extracted RFs than the inter-modality repeatability agreements. Similar patterns of repeatability agreements that exist across imaging modalities were also observed across HNC subtypes, as shown in Fig. 3d. Shape, unfiltered, and LoG filtered RFs had the highest repeatability agreement (MAD < 0.05, consistency ≥ 0.87), followed by wavelet-LLL/-HLL/-LHL (MAD: 0.02–0.05, consistency: 0.83–0.96). RFs from HHL-/**H-wavelet-filtered images showed the lowest repeatability agreement in terms of MAD (≥0.1) and consistency (0.69–0.83). Meanwhile, a significant proportion of RFs within the five wavelet-filtered categories had low repeatability (73–80% for NPC cohort and 70% for OPC cohort). Of note, 30% of all the extracted RFs (383/1288) with low repeatability were commonly found across the CECT images of the two HNC subtypes.

3.3   Feature Collinearity

The number and ratios of low/high-repeatable feature between the independent and all the extracted features are listed in Table 1. Notably, the differences in low-repeatable feature fractions are 0.06 maximum among all the four image sets. The portion of low/high repeatable features underwent minimum changes (maximum absolute difference = 0.06) after excluding all the volume-correlated features, as demonstrated in Fig. 4. The proportion of RFs with high volume correlation and high repeatability fluctuates between 0.06 and 0.1 for all the feature categories except shape features.

Table 1. Comparison of high repeatable feature counts and ratios between all the extracted features and independent features after clustering.

	All		Independent	
	Count	Ratio	Count	Ratio
NPC, CECT	836	0.65	163	0.66
NPC, CET1-w MR	870	0.67	172	0.66
NPC, T2-w MR	849	0.66	182	0.70
OPC, CECT	840	0.65	156	0.59

4   Discussion

Results of our study suggested that the majority of the shape, unfiltered, and LoG-filtered RFs demonstrated high repeatability (ICC ≥ 0.9) in all the studied image modalities and HNC subtypes. Notwithstanding, over 50% of the wavelet-filtered RFs exhibited weak repeatability, irrespective of image modalities and HNC subtypes. Notably, we observed numerous interesting fashions within the wavelet-filtered category. One example can be visually perceived in Fig. 2, where images with high-pass filtering on more dimensions demonstrated decreased feature repeatability. Specifically, wavelet-HHH and wavelet-LLL expressed an overwhelming disparity in RF repeatability.

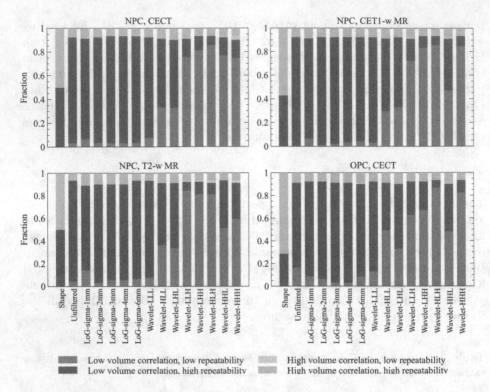

Fig. 4. Stacked bar plots containing the fractions of high-low volume correlated and high-/low repeatable radiomics features in each feature category for the four studied image sets.

The lower repeatability of RFs from wavelet-filtered images and their distinct patterns could partially be ascribed to the principle of the wavelet filter. A high-pass wavelet filter collects noisy and sharp edge signals, while a low-pass filter smooths the images. Hence, high-pass filtering could result in a more heterogeneous distribution of pixel values along the applied axis. Our perturbation algorithm alters two axes directions, which might elucidate our observation that the more dimensions the high-pass-filter applies to, the fewer repeatable RFs remain, and that HHH-wavelet RFs had the worst performance.

Our data demonstrated high repeatability agreements in all the compared image sets in general (MAD ≤ 0.06, consistency ≥ 0.89). The marginally drop in the agreement between cancer subtypes, compared to the inter-modality agreement, might be attributed to the discrepancies in ROI volumes. The OPC cohort has smaller ROI volumes (Fig. 5). The higher surface-to-volume ratios caused by smaller volumes led to larger relative variations of image intensity distributions within ROIs, contributing to the decreased RF repeatability under the applied perturbations. Herein, we speculate the impact of the rigid perturbations on RF repeatability might, to a large extent, depend directly on image filters and the inherent image characteristics such as ROI volume, rather than on the types of image modalities and cancer subtypes. Nevertheless, there might be other contributing factors and are worthy of further investigation.

In light of the progressively increasing adoption of wavelet-filter within the radiomics community in recent years [15, 19], our scrutiny of category-based RF repeatability is of paramount importance for RF pre-selection and robust model construction. Of note, various studies reported that over 90% of the key features in their models originated from wavelet-filtered images [15, 19]. However, among the selected wavelet RFs reported in the literature for HNC cases in MR images, we only observed high repeatability in 17/36 RFs, while certain extremely under-performing RFs (ICC $\leq$ 0.5) were noted (Table 4). Although our study inten-tionally focused on revealing positional variation dependence of RF repeatability, we, herein, argue that an even larger proportion of the underperforming (espe-cially wavelet-filtered) RFs would likely be foreseen when additional factors come into play. Thereby, we stress our pressing concerns on cautious handling of the wavelet-filtered RFs within the radiomics community.

Our study has limitations that need to be addressed in future studies. First, the perturbation algorithm might not fully mimic the positional variations as in real clinical scenarios owing to technical challenges in simulating small defor-mations of the patient's body between positionings. Second, although minimum bias was found from feature collinarity, which was consistent with the conclu-sion addressed by Fiset et al. [5], there are other potential confounding factors including image filtering settings, bin size/bin counts in image discretization, re-segmentation range, and radiomics calculation software. Third, a number of works were not accomplished in our research for the sake of maintaining compre-hensiveness while minimizing complexity. This includes investigating the impli-cation of RF repeatability to predictive model building and the agreement of our RF repeatability results in different cancer types or in a phantom study. We encouraged the community to carry out further investigations and will consider an extension of this work in the future.

5   Conclusions

In conclusion, although most RFs from unfiltered and LoG-filtered images demonstrated high repeatability, more than half of the wavelet-filtered RFs had poor repeatability, regardless of imaging modalities and HNC subtypes. Besides, RF repeatability agreements between imaging modalities were outstand-ing, while slightly worse between cancer subtypes. Minimum bias was observed from feature collinarity. Herein, we urge caution when handling wavelet-filtered RFs and advise excluding underperforming RFs during feature pre-selection for robust model construction.

6 Appendix

Table 2. Image preprocessing and feature extraction parameters.

	CECT	CET1-w MR, T2-w MR
N4B bias correction maximum iterations	N/A	[50, 40, 30]
Normalization reference structure	N/A	Brainstem
Normalization rescale factor	N/A	25
Pixel value offset	2000	2000
Resample pixel size (mm)	[1, 1, 1]	[1, 1, 1]
Anti-aliasing low-pass filter	Gaussian, $\beta = 0.97$	Gaussian, $\beta = 0.97$
Image/mask interpolation algorithm	Trilinear	Trilinear
CT image intensity rounding	No	N/A
Mask partial volume threshold	0.5	0.5
Interpolation grid alignment	Align grid origins	Align grid origins
Image thresholding	$\pm 3\sigma$	$\pm 3\sigma$
Translation distances (pixel)	[0.0, 0.2, 0.4, 0.6, 0.8]	[0.0, 0.2, 0.4, 0.6, 0.8]
Rotation angles (degree)	[−20, −15, −10, −5, 0, 5, 10, 15, 20]	[−20, −15, −10, −5, 0, 5, 10, 15, 20]
Image discretization bin size	10	10
Image filters	Unfiltered, Laplacian-of-Gaussian, Wavelet	Unfiltered, Laplacian-of-Gaussian, Wavelet
Kernel size of Laplacian-of-Gaussian filter (mm)	[1, 2, 3, 4, 6]	[1, 2, 3, 4, 6]
Wavelet filter starting level	0	0
Wavelet filter total level	1	1
Wavelet filter type	Coilf1	Coilf1
Wavelet filter decompositions	[LLL, HLL, LHL, LLH, LHH, HLH, HHL, HHH]	[LLL, HLL, LHL, LLH, LHH, HLH, HHL, HHH]
Feature class	Shape, firstorder, glcm, glrlm, glszm, gldm, ngtdm	Shape, firstorder, glcm, glrlm, glszm, gldm, ngtdm

Table 3. Distribution of low repeatability radiomics feature across different imaging modalities and HNC subtypes

Tumor subtype			NPC			OPC
Image modality			CET1-w MR	T2-w MR	CECT	CECT
Low repeatable features (ICC > 0.9)	Shape		0%	0%	0%	0%
	Unfiltered		0%	4.4%	3.3%	16.5%
	LoG filtered		3.5%	7.0%	4.0%	6.4%
	Wavelet filtered	LLL	3.3%	7.7%	7.7%	13.2%
		HLL, LHL	31.3%	35.2%	33.0%	41.2%
		HHL, **H	75.2%	73.0%	80.0%	69.7%
		All wavelet	55.2%	55.4%	59.2%	55.5%
	All		32.4%	34.1%	35.1%	34.8%
Commonly low-repeatable features			28.5%			N/A
			29.7%		29.7%	

Table 4. Repeatability of wavelet feature used as final selected features in previous literature. Feature repeatabibilty was quantified as intra-class correlation coefficient (ICC). Low repeatable RFs were marked by star (*).

Feature (ICC)	Image modality
HLL-wavelet Category	
HLL-first-order-median (0.94)	CET1-w MR
HLL-glrlm-run-percentage (0.98)	CET1-w MR
HLL-glcm-correlation (0.90)	CET1-w MR
HLL-glcm-cluster-prominence (0.95)	T2-w MR
HLL-gldm-dependence-entropy (0.98)	T2-w MR
HLL-gldm-small-dependence-low-gray-level-emphasis (0.61)	T2-w MR
HLL-ngtdm-complexity (0.90)	T2-w MR
LHH-wavelet Category	
LHH-first-order-mean (0.92)	CET1-w MR
LHH-first-order-mean (0.90)	T2-w MR
LHH-first-order-median (0.75)	CET1-w MR

(*continued*)

Table 4. (*continued*)

Feature (ICC)	Image modality
LHH-glszm-gray-level-non-uniformity-normalized (0.30)*	CET1-w MR
LHH-glszm-small-area-high-gray-level-emphasis (0.50)*	CET1-w MR
LLH-wavelet Category	
LLH-first-order-mean (0.98)	CET1-w MR
LLH-first-order-median (0.93)	CET1-w MR
LLH-first-order-mean (0.99)	T2-w MR
LLL-wavelet Category	
LLL-glcm-cluster-shade (1.00)	CET1-w MR, T2-w MR
LLL-glcm-inverse-variance (1.00)	T2-w MR
LLL-glrlm-short-run-low-gray-level-emphasis (0.89)	T2-w MR
LLL-glrlm-long-run-high-gray-level-emphasis (0.99)	T2-w MR
HHL-wavelet Category	
HHL-glszm-zone-size-non-uniformity-normalized (0.67)	CET1-w MR
HHL-first-order-mean (0.62)	T2-w MR
HHL-glcm-sum-average (0.70)	T2-w MR
HLH-wavelet Category	
HLH-first-order-skewness (0.70)	CET1-w MR
HLH-glcm-informational-measure-of-correlation-1 (0.66)	CET1-w MR
HLH-glcm-informational-measure-of-correlation-1 (0.62)	T2-w MR
HLH-glcm-informational-measure-of-correlation-2 (0.62)	T2-w MR
HLH-first-order-rms (0.86)	T2-w MR
HLH-glcm-autocorrelation (0.45)*	T2-w MR
LLH-wavelet Category	
LLH-glrlm-long-run-high-gray-level-emphasis (0.66)	CET1-w MR
LLH-first-order-skewness (0.90)	CET1-w MR
LLH-glcm-cluster-shade (0.89)	T2-w MR
LLH-glcm-correlation (0.63)	T2-w MR
LLH-ngtdm-strength (0.75)	T2-w MR
LLH-glszm-size-zone-non-uniformity-normalized (0.17)*	T2-w MR
LHL-wavelet Category	
LHL-glszm-small-area-high-gray-level-emphasis (0.86)	CET1-w MR

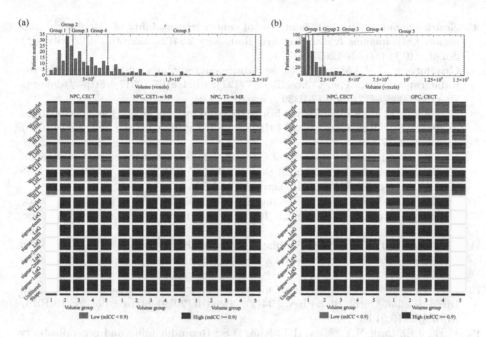

Fig. 5. Category-based binary radiomics feature repeatability separated by volume groups for (a) CECT, CET1-w MR, and T2-w MR of the NPC cohort and (b) CECT of the NPC cohort and CECT of the OPC cohort. The top figure is the histogram of the ROI volume for the NPC patient cohort, and the dashed black lines indicate the four threshold values for patient grouping.

References

1. Beare, R., Lowekamp, B., Yaniv, Z.: Image segmentation, registration and characterization in R with SimpleiTK. J. Stat. Softw. **86**(1), 1–35 (2018). https://doi.org/10.18637/jss.v086.i08

2. Bianchini, L., et al.: A multicenter study on radiomic features from T2-weighted images of a customized MR pelvic phantom setting the basis for robust radiomic models in clinics. Magn. Reson. Med. **00**, 1–14 (2020). https://doi.org/10.1002/mrm.28521

3. Dehing-Oberije, C., et al.: Tumor volume combined with number of positive lymph node stations is a more important prognostic factor than TNM stage for survival of non-small-cell lung cancer patients treated with (chemo)radiotherapy. Int. J. Radiat. Oncol. Biol. Phys. **70**(4), 1039–1044 (2008). https://doi.org/10.1016/j.ijrobp.2007.07.2323

4. Elshafeey, N., et al.: Multicenter study demonstrates radiomic features derived from magnetic resonance perfusion images identify pseudoprogression in glioblastoma. Nat. Commun. **10**(1), 3170 (2019). https://doi.org/10.1038/s41467-019-11007-0

5. Fiset, S., et al.: Repeatability and reproducibility of MRI-based radiomic features in cervical cancer. Radiother. Oncol. **135**, 107–114 (2019). https://doi.org/10.1016/j.radonc.2019.03.001

6. Gourtsoyianni, S., et al.: Primary rectal cancer: repeatability of global and local-regional MR imaging texture features. Radiology **284**(2), 552–561 (2017). https://doi.org/10.1148/radiol.2017161375
7. Griethuysen, V.J.J.M., et al.: Computational radiomics system to decode the radiographic phenotype. Cancer Res. **77**(21), e104–e107 (2017). https://doi.org/10.1158/0008-5472.CAN-17-0339
8. Lambin, P., et al.: Radiomics: the bridge between medical imaging and personalized medicine. Nat. Rev. Clin. Oncol. **14**(12), 749–762 (2017). https://doi.org/10.1038/nrclinonc.2017.141
9. Liu, R., et al.: Stability analysis of CT radiomic features with respect to segmentation variation in oropharyngeal cancer. Clin. Transl. Radiat. Oncol. **21**, 11–18 (2020). https://doi.org/10.1016/j.ctro.2019.11.005
10. Lu, H., et al.: Repeatability of quantitative imaging features in prostate magnetic resonance imaging. Front. Oncol. **10**, 551 (2020). https://doi.org/10.3389/fonc.2020.00551
11. McGraw, K.O., Wong, S.P.: Forming inferences about some intraclass correlation coefficients. Psychol. Methods **1**(1), 30–46 (1996). https://doi.org/10.1037/1082-989X.1.1.30
12. Nie, K., et al.: NCTN assessment on current applications of radiomics in oncology. Int. J. Radiat. Oncol. Biol. Phys. **104**(2), 302–315 (2019). https://doi.org/10.1016/j.ijrobp.2019.01.087
13. Park, J.E., Park, S.Y., Kim, H.J., Kim, H.S.: Reproducibility and generalizability in radiomics modeling: possible strategies in radiologic and statistical perspectives. Korean J. Radiol. **20**(7), 1124–1137 (2019). https://doi.org/10.3348/kjr.2018.0070
14. Schwier, M., et al.: Repeatability of multiparametric prostate MRI radiomics features. Sci. Rep. **9**(1), 9441 (2019). https://doi.org/10.1038/s41598-019-45766-z
15. Sheikh, K., et al.: Predicting acute radiation induced xerostomia in head and neck cancer using MR and CT radiomics of parotid and submandibular glands. Radiat. Oncol. **14**(1), 131 (2019). https://doi.org/10.1186/s13014-019-1339-4
16. Traverso, A., Wee, L., Dekker, A., Gillies, R.: Repeatability and reproducibility of radiomic features: a systematic review. Int. J. Radiat. Oncol. Biol. Phys. **102**(4), 1143–1158 (2018). https://doi.org/10.1016/j.ijrobp.2018.05.053
17. van Velden, F.H.P., et al.: Repeatability of radiomic features in non-small-cell lung cancer [18F]FDG-PET/CT studies: impact of reconstruction and delineation. Mol. Imag. Biol. **18**(5), 788–795 (2016). https://doi.org/10.1007/s11307-016-0940-2
18. Wang, G., He, L., Yuan, C., Huang, Y., Liu, Z., Liang, C.: Pretreatment MR imaging radiomics signatures for response prediction to induction chemotherapy in patients with nasopharyngeal carcinoma. Eur. J. Radiol. **98**, 100–106 (2018). https://doi.org/10.1016/j.ejrad.2017.11.007
19. Zhang, L., et al.: Radiomic nomogram: pretreatment evaluation of local recurrence in nasopharyngeal carcinoma based on MR imaging. J. Cancer **10**(18), 4217–4225 (2019). https://doi.org/10.7150/jca.33345
20. Zhang, L.L., et al.: Pretreatment MRI radiomics analysis allows for reliable prediction of local recurrence in non-metastatic T4 nasopharyngeal carcinoma. EBioMedicine **42**, 270–280 (2019). https://doi.org/10.1016/j.ebiom.2019.03.050
21. Zwanenburg, A., et al.: Assessing robustness of radiomic features by image perturbation. Sci. Rep. **9**(1), 614 (2019). https://doi.org/10.1038/s41598-018-36938-4

MLCN: Metric Learning Constrained Network for Whole Slide Image Classification with Bilinear Gated Attention Mechanism

Baorong Shi, Xinyu Liu, and Fa Zhang$^{(\boxtimes)}$

Institute of Computing Technology, Chinese Academy of Science, Beijing, China
{shibaorong19s,liuxinyu,zhangfa}@ict.ac.cn

Abstract. Whole Slide Image (WSI) classification is an important part of pathological diagnosis. Although previous approaches (such as DSMIL and CLAM) have achieved good results, the classification performance is still unsatisfactory because the learned features of WSI lack discrimination and the correlation among sub-characteristics of tumor images are ignored. In this paper, we proposed a Metric Learning Constraint Network (referred to as MLCN). Particularly, MLCN benefits from two aspects: 1) It enhances the discriminative power of features by enlarging inter-class distance and narrowing intra-class distance in both slide-level and patch-level. 2) It learns a more powerful feature aggregator by proposing the bilinear gated attention mechanism to capture relations among sub-characteristics of tumor issues. Experiments on CAMELYON16 and TCGA Kidney datasets validate the effectiveness of our approach, and we achieved state-of-the-art performance compared to other popular methods. The codes will be available soon.

Keywords: Deep learning · Pathological image · Whole Slide Image classification · Metric learning · Attention mechanism

1 Introduction

Histopathology tissue analysis plays a critical role in cancer diagnosis and prognosis. Particularly, pathological images classification is the core task in histopathology tissue analysis.

Traditionally, doctors classify pathological images with the naked eye. With the development of technology, Computer Aid Diagnosis (CAD) has become popular. As the most widely used pathological image in CAD, WSI (Whole Slide Image) is the digital scanning results of histopathology tissues [12,13]. Since a WSI contains millions of cells, is of high resolution and occupies a great amount of memory, it becomes impossible to directly analyse it. Generally, a WSI is divided into many patches for subsequent process. In this context, WSI classification includes two types: strongly supervised learning and weakly supervised

© The Author(s), under exclusive license to Springer Nature Switzerland AG 2022
W. Qin et al. (Eds.): CMMCA 2022, LNCS 13574, pp. 35–46, 2022.
https://doi.org/10.1007/978-3-031-17266-3_4

learning. Strongly supervised learning means that each patch in WSI has its own label, while weakly supervised learning refers to that all patches in WSI only have a slide-level label. In practice, patch-level annotations of WSIs are too hard to obtain because it requires a lot of labor and great professionalism. Thus researchers pay more attention to weakly supervised learning, which models WSI classification as a MIL (Multi-Instance Learning) problem [23]. In traditional machine learning-based MIL, many works, such as Citation-kNN [26], EM-DD [27], MI-Kernels [28] and so on, are proposed. But these methods have the possibility of high errors, need complex algorithm selection and consume large amounts of time and space. Consequently, more researchers pay attention to deep learning-based MIL approaches [17,18,31], which usually comprise two stages: learning representations and representations aggregation. Learning representations encodes patches of a WSI into patch-level embeddings or scores with CNNs, while representation aggregation summarizes the patch-level embeddings or scores to generate the slide-level embedding or score using MIL pooling functions, which includes max-pooling [14], mean-pooling [15], Noisy-AND [24], attention mechanism [1] and so on. Recently, many deep learning-based MIL works have achieved good results, such as [1,2,4,5]. However, there still exist two challenges which limit the performance of WSI classification.

Firstly, the learned representations lack discrimination. Creating a WSI needs multiple processes [19]: slicing, placing, staining and scanning, and it's easy to introduce unrelated noises, such as wrinkles, blur and color variance of tissue slices. Thus, pathological images are more likely to have large differences within the same category and small differences between different categories. In response to it, some machine learning-based works use metrics like bag distances or bag similarities to enforce intra-class compactness and inter-class discrepancy [29]. However, there is no research to enhance feature discrimination with deep learning [29]. Since poor feature discrimination will directly affect the classification accuracy, how to enlarge inter-class distance and minimize intra-class distance with deep learning becomes an urgent topic for WSI classification.

Secondly, existing attention mechanisms in representations aggregation are not powerful enough. It is known that some sub-characteristics of tumor images, such as color, texture, shape and size, are not independent of each other in clinical diagnosis. And there is specific correlation among these sub-characteristics, which is of great importance for WSI analysis. Although attention-based aggregators are proved to be the most effective in MIL pooling, existing attention mechanisms are still too trivial to capture the correlation among sub-characteristics of tumor images, which prevents the model from learning richer information about WSIs and undoubtedly limits the classification performance.

To address these challenges, we proposed a novel Metric Learning Constraint Network (refered to as MLCN). MLCN uses learned attention to aggregate the patch embeddings extracted by encoder to get slide-level features, which are then used for label prediction. Particularly, the main innovations of MLCN lie in improving the attention mechanism and enhancing the discrimination of features. To improve the attention mechanism, we model the correlation among sub-characteristics of tumor images with bilinear operation, which is inspired by [6]. Then, it is combined with the gated attention mechanism [1] to learn more powerful attention and aggregate features more effectively. And we name the new attention mechanism as bilinear gated attention mechanism. To enhance the discrimination of features, we proposed center cluster loss, which is inspired by some typical metric learning-based works [9,10,20]. Learning a center feature for each class of WSIs, center cluster loss not only penalizes distances between the learned slide-level features and their corresponding center features, but also penalizes the relative distances among center features, patch-level features of key instances and patch-level features of non-key instances. In the course of training, the loss could enforce intra-class compactness and inter-class discrepancy in both slide-level and patch-level. More details about MLCN will be illustrated in the Method section. Experiment results on two public datasets show that our MLCN outperforms other popular methods has achieved state-of-the-art performance.

2 Method

MLCN contains two key innovations: bilinear gated attention mechanism and center cluster loss. Specially, it uses bilinear gated attention mechanism to aggregate patch-level embeddings extracted by pretrained CNN model to get slide-level embeddings, then center cluster loss is used to constrain feature space in both patch-level and slide-level. We'll first introduce bilinear gated attention and illustrate center cluster loss, then illustrate our MLCN from a global perspective.

2.1 Bilinear Gated Attention Mechanism

Attention mechanism is an important MIL pooling approach. Let $H = \{h_1, h_2, ..., h_N\}$, where H is a bag that contains all the patch embeddings of a WSI, and $h_i \in R^{1 \times D}, i \in \{1, 2, ..., N\}$ is the embedding of patch i. Then attention-based MIL pooling is given by

$$Z = \sum_{i=1}^{N} a_i h_i \qquad (1)$$

In the state-of-the-art attention mechanism, Gated attention mechanism [1], a_i is given by

$$a_i = \frac{exp\{w^T(tanh(Vh_i^T) \odot sigm(Uh_i^T))\}}{\sum_{j=1}^{N} exp\{w^T(tanh(Vh_j^T) \odot sigm(Uh_j^T))\}} \qquad (2)$$

where $V \in R^{C \times D}$, $U \in R^{C \times D}$, and $w \in R^{C \times 1}$ are linear projection layers. $\odot$ is element-wise multiplication. According to [1], the output of $tanh(\cdot)$ contains both positive and negative values, which prompts proper gradient flow. And sigmoid activation function is introduced to bring more non-linearity in $[-1, 1]$ since $tanh(x)$ is nearly linear for $x \in [-1, 1]$. The gated attention learns from the gated mechanism in [22] and captures similarities among instances. And it is widely used in many works [2, 30].

However, the gated attention mechanism only focuses on learning relations among instances but ignores the correlation of sub-characteristics. It is known that there is specific correlation among these sub-characteristics of tumor images, such as color, texture, shape and so on. The correlation is also an critical factor in pathological diagnosis. Consequently, capturing the correlation of sub-characteristics would benefit WSI classification.

We proposed bilinear gated attention mechanism to model the correlation. Let $t_i = tanh(Vh_i)$, $t_i \in R^{1 \times C}$ and it represents the output feature of patch i produced by linear projection V and $tanh(\cdot)$; $s_i = sigm(Uh_i)$, $s_i \in R^{1 \times C}$ and it represents the output feature of patch i produced by linear projection U and $sigm(\cdot)$. Each feature channel in t_i and s_i denotes a sub-characteristic. Then the pairwise interactions among sub-characteristics are given by

$$X = \sum_{i=1}^{N} t_i^T s_i \tag{3}$$

X is a $C \times C$ matrix. The formulation models the correlation among sub-characteristics of tumor patches by calculate cosine similarity. And it is a bilinear operation since there are two factors and the output of the equation is linear in either factor when the other is held constant. Then softmax operation is applied on X to get X'. Elements of each row in X' are limited to (0,1) and the sum of them equals to 1, which is given by

$$X' = softmax(X, dim = 1) \tag{4}$$

X' is then used as a score matrix to fuse with the gated attention mechanism to learn more powerful attention and help aggregate features more effectively, which is given by

$$a_i = \frac{exp\{w^T((tanh(Vh_i^T) \odot sigm(Uh_i^T))X')\}}{\sum_{j=1}^{N} exp\{w^T((tanh(Vh_j^T) \odot sigm(Uh_j^T)X'))\}} \tag{5}$$

2.2 Center Cluster Loss

There are two kinds of features in WSI classification: patch-level features and slide-level features. Since slide-level features are aggregated with patch-level features, the discrimination of patch-level features would affect the discriminative power of slide-level features, while the discrimination of slide-level features is closely related to classification performance. Thus we need to take into account both slide-level and patch-level features when enhancing discriminative power. Moreover, an important prior is that the label of key instances should be consistent with the label of its slide. In this context, we proposed center cluster loss. It learns a center feature x_k for each class k of WSIs, $k \in \{1, 2, ..., K\}$ and contains two kinds of constraints: slide-level constraint and patch-level constraint.

Slide-level constraint prompts the learned slide-level feature to move closer to its corresponding class center. Let f_s represents the slide-level feature of a WSI, which belongs to class k. Then the constraint is given by

$$L_s = \frac{1}{2}||f_s - x_k||^2 \tag{6}$$

It penalizes distances between slide-level features and their corresponding center features of class k. Relatively, the inter-class distances are enlarging in the course of training, which enhances discriminative power of slide-level features.

Patch-level constraint penalizes the relative distances among center features, patch-level features of key instances and patch-level features of non-key instances. Let $f_p = \{f_{p1}, f_{p2}, ..., f_{pN}\}$ denote the set of patch-level features of a WSI, which belongs to class k. According to the attention score learned by bilinear gated attention mechanism, we select top m instances with highest scores and select bottom m instances with lowest scores. Among the top m instances, we choose the instance that is most distant from the center feature of class k as the positive instance, its features named as f_{pp}. Among the bottom m instances, we choose the instance that is closest to the center feature of class k as the negative instance, its features named as f_{pn}. And the patch-level constraint is given by

$$L_p = log(1 + exp\{||f_{pp} - x_k||^2 - ||f_{pn} - x_k||^2\}) \tag{7}$$

Through hard example mining and penalizing relative distances, the constraint prompts the patch-level features of key instances closer to center feature x_k and prompts the patch-level features of non-key instances more distant from x_k. Since the slide-level feature of a WSI is aggregated with all the patch-level features and center feature x_k is the ideal average of slide-level features, L_p constraint increases the contribution of key instances and descreases the contribution of non-key instances to the slide-level feature. What's more, it enlarges the distances between key instances and non-key instances, which enhances the discrimination of patch-level features.

Finally, center cluster loss is given by

$$L_c = L_s + L_p \tag{8}$$

Center cluster loss enhances the discriminative power of features in both slide-level and patch-level, which strengthens the model's ability to understand WSIs and contributes to the performance of WSI classification.

2.3 A Global Perspective of MLCN

We integrated the two innovations into MLCN. Notice that as a loss that constrains features learning, center cluster loss is a part of total loss. To introduce more priors to constrain feature space, we also apply the Instance-level Clustering, which is firstly proposed in CLAM [2], into our MLCN and we name it L_i.

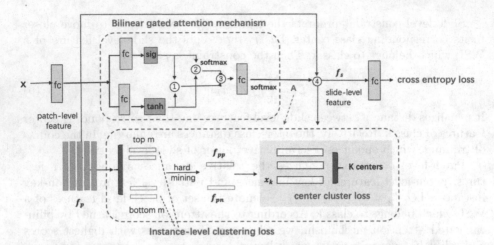

Fig. 1. Metric Learning Constraint Network: First, X, the embeddings of all patches in a WSI, which are extracted through pretrained ResNet50, are linearly projected to obtain patch-level features f_p. Second, the patch-level features are aggregated by bilinear gated attention mechanism to obtain slide-level features f_s. And A is the learned attention score. Third, classification loss, center cluster loss and instance-level clustering loss are computed based on f_s, f_p and A to train the whole network. Notice that ①,②,③,④ are corresponding operations mentioned in Sect. 2.1.

In addition, cross-entropy loss is used as a basic classification loss, and we name it L_b. The total loss of MLCN is given by

$$L = L_b + \lambda L_c + \gamma L_i \tag{9}$$

where λ and γ are balance factor. In a nutshell, MLCN is shown in Fig. 1, and the details of it are illustrated in the caption. More details about Instance-level Clustering [2] is shown in Fig. 2.

3   Experiments

3.1   Experimental Settings

WSI Processing. WSI processing contains two stages. The first stage is to cut a WSI into patches. To avoid cutting blank areas, we apply the automated segmentation algorithm in [2] on WSIs after they are read at a 64× downsampled resolution. Then we crop 256 × 256 patches from the segmented foreground areas and stack them according to their coordinates. The second stage is to extract features for patches. A Resnet-50 model [8] pretrained on ImageNet [11] is used here. And we take the output of block 3 and perform adaptive mean-spatial pooling on it to generate a 1024-dim embedding for each patch. These patch embeddings of a WSI serve as the input to MLCN.

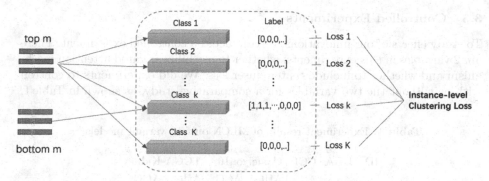

Fig. 2. Instance-level Clustering: The input is the concatenation of top m and bottom m patch-level features selected in Sect. 2.2. A prior introduced here is that cancer subtypes are mutually exclusive. It assumes that the labels of top-m instances are the same as its slide-level label. We learn K fully-connected layers, which represents K classes. Each FC layer projects the input to a 2m-length logits. Each loss is cross entropy loss and Instance-level clustering loss is the sum of these losses.

Implementation Details. When training MLCN and validating its performance, we use 10-fold monte carlo cross-validation. For each fold, we randomly split the dataset into a training set (75%), a validation set (15%) and a test set (10%). And we use the training set to train MLCN, with Adam [16] optimizer, batch size set as 1, initial learning rate set as 0.0001 and weight decay set as 0.00001. For each fold, the classification performance (AUC and ACC) on validation set is used to filter the best model with the stopping criterion being that the number of training epochs exceeds 50 and the validation loss has not decreased for 20 consecutive epochs. Moreover, the λ and γ in center cluster loss are separately set as 0.4 and 0.003 by our study. The m in hard example mining is set as 8. C and D in Sect. 2.1 are seperately set as 256 and 512. The experiments are performed on NVIDIA GeForce RTX 2080 Ti GPU.

3.2   Dataset

Camelyon16. Camelyon16 [3] is a public breast cancer lymph node metastasis detection dataset, which includes two class: normal and tumor. It contains 271 images in the train set, with 160 of them normal and 111 of them tumor. It also contains 129 images in the test set.

TCGA Kidney. TCGA Kidney dataset is made of kidney images from TCGA projects [25], including TCGA-KICH (Kidney Chromophobe), TCGA-KIRC (Kidney Renal Clear Cell Carcinoma), TCGA-KIRP (Kidney Renal Papillary Cell Carcinoma). It contains 2068 images with 326 of them from TCGA KICH, 994 of them from KIRP and 748 of them from KIRC, in which only the slide-level label is available. It is a 3-class dataset.

3.3   Controlled Experiments

To study effects of our innovations, we conducted a controlled experiment. There are 2 variables in this experiments: whether to use bilinear gated attention mechanism and whether to include center cluser loss. We did experiments to controll with or without the two variables for a comparative study, as shown in Table 1.

Table 1. Experiment results of MLCN on four variant models.

ID	BGA	CCL	Camelyon16		TCGA-Kidney	
			AUC	ACC	AUC	ACC
M1	✗	✗	0.884	0.840	0.979	0.914
M2	✓	✗	0.904	0.864	0.981	0.917
M3	✗	✓	0.913	0.870	0.98	0.920
M4	✓	✓	**0.914**	**0.878**	**0.982**	**0.921**

In this table, BGA represents bilinear gated attention mechanism and CCL represents center cluster loss. Then, M1 indicates the model without BGA and CCL. Compared with M4, the attention in M1 is equation (2) and its loss is $L_b + \gamma L_i$. M2 indicates the model with only BGA. Compared with M4, the attention in M2 is equation (2). M3 indicates the model with only CCL. Compared with M4, the loss in M3 is $L_b + \gamma L_i$. M4 indicates our MLCN with both BGA and CCL. From the table, the comparison between M2 and M1 validates that bilinear gated attention mechanism could help learn the correlation among sub-characteristics of tumor images. The comparison between M3 and M1 validates the efficacy of center cluster loss to enhance the discriminative power of features. M4 outperforms the other three models on two datasets with 0.914 in AUC, 0.878 in ACC on Camelyon16 and 0.982 in AUC, 0.921 in ACC on TCGA Kidney, which proves that the combination of BGA and CCL increased the model's ability to better understand WSIs.

Table 2. Experiment results for MIL aggregators

MIL aggregators	Camelyon16		TCGA-Kidney	
	AUC	ACC	AUC	ACC
Mean-pooling [15]	0.724	0.694	0.783	0.802
Max-pooing [14]	0.911	0.854	0.977	0.912
Standard attention mechanism [1]	0.906	0.862	0.98	0.918
Gated attention mechanism [1]	0.913	0.870	0.98	0.920
Bilinear gated attention mechanism	**0.914**	**0.878**	**0.982**	**0.921**

3.4   Ablation Study for MIL Aggregators

To validate the effectiveness of bilinear gated attention mechanism, we compared different MIL pooling approaches. In experiments, we only changed the MIL pooling method and kept other settings the same. Experiment results are shown in Table 2. The first two rows of the table are traditional MIL aggregators: mean-pooling and max-pooling. The third row and fourth row are two recent attention-based MIL pooling methods. The last row is our method. From the table, we learn that our bilinear gated attention mechanism outperforms other four approaches and has achieved better performance.

Table 3. Experiment results for different losses

Loss	Camelyon16		TCGA-Kidney	
	AUC	ACC	AUC	ACC
L_b	0.898	0.852	0.979	0.909
$L_b + L_i$	0.904	0.864	0.981	0.917
$L_b + L_i + L_c$	**0.914**	**0.878**	**0.982**	**0.921**

3.5   Ablation Study for Loss

To find out the effects of different components in the total loss, we performed ablation study. We kept other settings the same but changed the loss function in this experiment. In this table, L_b represents the basic classification loss: cross entropy loss, L_i represents Instance-level Clustering Loss, and L_c represents Center Cluster loss, which are illustrated in Sect. 2.2. From the Table 3, we learn that the combination of L_b and L_i are better than only L_b, and the combination of L_b, L_i, L_c are better than the combination of L_b, L_i.

3.6   Compare with State-of-the-Art Approahces

We compare MLCN with other SOTA deep MIL models, including MILRNN [4], ABMIL [1], CLAM [2] and DSMIL [5]. Notice that for approaches designed for two-class MIL problem, we calculate the final score for each class seperately to handle multi-class problems on TCGA Kidney. From the Table 4, we learn that MLCN outperforms other four approaches, with at least about 1.6%, at most about 9.3% improvement. Compared with other approaches, MLCN learns the correlation among sub-characteristics of tumor images by using bilinear gated attention, and it enhances the discrimination of features by using center cluster loss to constrain feature space. These innovations help boost the WSI classification.

Table 4. Comparison with other state-of-the-art methods

Methods	Camelyon16		TCGA-Kidney	
	AUC	ACC	AUC	ACC
MILRNN [4]	0.806	0.806	0.875	0.835
ABMIL [1]	0.865	0.845	0.923	0.889
CLAM [2]	0.884	0.840	0.979	0.914
DSMIL [5]	0.894	0.868	0.963	0.907
MLCN	**0.914**	**0.878**	**0.982**	**0.921**

4   Conclusion

In this paper, we proposed a metric learning constraint network with bilinear gated attention mechanism, which not only captures the correlation among sub-characteristics of tumor images but also enhances the discriminative power of features in both slide-level and patch-level. Experiments on Camelyon16 dataset and TCGA Kidney dataset validated the effectiveness of our innovations. Surprisingly, our MLCN outperforms other popular methods and has achieved state-of-the-art performance.

Acknowledgements. The research is supported by the Strategic Priority Research Program of the Chinese Academy of Sciences (No. XDA16021400), and the NSFC projects grants (61932018, 62072441 and 62072280).

References

1. Ilse, M., Tomczak, J., Welling, M.: Attention-based deep multiple instance learning. In: International Conference on Machine Learning, pp. 2127–2136 (2018)
2. Lu, M.Y., Williamson, D.F., Chen, T.Y., et al.: Data-efficient and weakly supervised computational pathology on whole-slide images. Nat. Biomed. Eng. **5**(6), 555–570 (2021)
3. Bejnordi, B.E., Veta, M., Van Diest, P.J., Van Ginneken, B., et al.: Diagnostic assessment of deep learning algorithms for detection of lymph node metastases in women with breast cancer. JAMA **318**(22), 2199–2210 (2017)
4. Campanella, G., Hanna, M.G., Geneslaw, L., et al.: Clinical-grade computational pathology using weakly supervised deep learning on whole slide images. Nat. Med. **25**(8), 1301–1309 (2019)
5. Li, B., Li, Y., Eliceiri, K.W.: Dual-stream multiple instance learning network for whole slide image classification with self-supervised contrastive learning. In: Proceedings of the IEEE/CVF Conference on Computer Vision and Pattern Recognition, pp. 14318–14328 (2021)
6. Lin, T.Y., RoyChowdhury, A., Maji, S.: Bilinear CNN models for fine-grained visual recognition. In: Proceedings of the IEEE International Conference on Computer Vision, pp. 1449–1457 (2015)

7. Carreira, J., Caseiro, R., Batista, J., Sminchisescu, C.: Semantic segmentation with second-order pooling. In: Fitzgibbon, A., Lazebnik, S., Perona, P., Sato, Y., Schmid, C. (eds.) ECCV 2012. LNCS, vol. 7578, pp. 430–443. Springer, Heidelberg (2012). https://doi.org/10.1007/978-3-642-33786-4_32
8. He, K., Zhang, X., Ren, S., Sun, J.: Deep residual learning for image recognition. In: Proceedings of the IEEE Conference on Computer Vision and Pattern Recognition, pp. 770–778 (2016)
9. Schroff, F., Kalenichenko, D., Philbin, J.: FaceNet: a unified embedding for face recognition and clustering. In: Proceedings of the IEEE Conference on Computer Vision and Pattern Recognition, pp. 815–823 (2015)
10. Wen, Y., Zhang, K., Li, Z., Qiao, Yu.: A discriminative feature learning approach for deep face recognition. In: Leibe, B., Matas, J., Sebe, N., Welling, M. (eds.) ECCV 2016. LNCS, vol. 9911, pp. 499–515. Springer, Cham (2016). https://doi.org/10.1007/978-3-319-46478-7_31
11. Russakovsky, O., et al.: ImageNet large scale visual recognition challenge. Int. J. Comput. Vision **115**(3), 211–252 (2015). https://doi.org/10.1007/s11263-015-0816-y
12. Cornish, T.C., Swapp, R.E., Kaplan, K.J.: Whole-slide imaging: routine pathologic diagnosis. Adv. Anat. Pathol. **19**(3), 152–159 (2012)
13. Pantanowitz, L., Valenstein, P.N., Evans, A.J., et al.: Review of the current state of whole slide imaging in pathology. J. Pathol. Inform. **2**(1), 36 (2011)
14. Feng, J., Zhou, Z.H.: Deep MIML network. In: Proceedings of the AAAI Conference on Artificial Intelligence, vol. 31, No. 1 (2017)
15. Pinheiro, P.O., Collobert, R.: From image-level to pixel-level labeling with convolutional networks. In: Proceedings of the IEEE Conference on Computer Vision and Pattern Recognition, pp. 1713–1721 (2015)
16. Kingma, D.P., Ba, J.: Adam: a method for stochastic optimization. arXiv preprint arXiv:1412.6980 (2014)
17. Wang, D., Khosla, A., Gargeya, R., et al.: Deep learning for identifying metastatic breast cancer. arXiv preprint arXiv:1606.05718 (2016)
18. Hashimoto, N., Fukushima, D., Koga, R., et al.: Multi-scale domain-adversarial multiple-instance CNN for cancer subtype classification with unannotated histopathological images. In: Proceedings of the IEEE/CVF Conference on Computer Vision and Pattern Recognition, pp. 3852–3861 (2020)
19. Komura, D., Ishikawa, S.: Machine learning methods for histopathological image analysis. Comput. Struct. Biotechnol. J. **16**, 34–42 (2018)
20. Kaya, M., Bilge, H.Ş: Deep metric learning: a survey. Symmetry **11**(9), 1066 (2019)
21. Cheplygina, V., Tax, D.M., Loog, M.: Multiple instance learning with bag dissimilarities. Pattern Recogn. **48**(1), 264–275 (2015)
22. Dauphin, Y.N., Fan, A., Auli, M., et al.: Language modeling with gated convolutional networks. In: International Conference on Machine Learning, pp. 933–941 (2017)
23. Wang, X., Yan, Y., Tang, P., et al.: Revisiting multiple instance neural networks. Pattern Recogn. **74**, 15–24 (2018)
24. Kraus, O.Z., Ba, J.L., Frey, B.J.: Classifying and segmenting microscopy images with deep multiple instance learning. Bioinformatics **32**(12), i52–i59 (2016)
25. Atlas, T.C.G. Website (2006). https://portal.gdc.cancer.gov/
26. Jun, W., Jean-Daniel, Z.: Solving the multiple-instance problem: a lazy learning approach. In: Proceedings of the 17th International Conference on Machine Learning, pp. 1119–1125 (2000)

27. Zhang, Q., Goldman, S.: EM-DD: an improved multiple-instance learning technique. Adv. Neural Inf. Process. Syst. **14** (2001)
28. Gärtner, T., Flach, P.A., Kowalczyk, A., et al.: Multi-instance kernels. In: ICML, vol. 2, p. 7 (2002)
29. Ilse, M., Tomczak, J.M., Welling, M.: Deep multiple instance learning for digital histopathology. In: Handbook of Medical Image Computing and Computer Assisted Intervention, pp. 521–546 (2020)
30. Han, Z., Wei, B., Hong, Y., et al.: Accurate screening of COVID-19 using attention-based deep 3D multiple instance learning. IEEE Trans. Med. Imaging **39**(8), 2584–2594 (2020)
31. Yan, R., Ren, F., Wang, Z., et al.: Breast cancer histopathological image classification using a hybrid deep neural network. Methods **173**(1), 52–60 (2020)

NucDETR: End-to-End Transformer for Nucleus Detection in Histopathology Images

Ahmad Obeid(✉) [iD], Taslim Mahbub, Sajid Javed[iD], Jorge Dias[iD], and Naoufel Werghi[iD]

Khalifa University, Abu Dhabi, UAE
ahmad.obeid@ku.ac.ae

Abstract. Nucleus detection in histopathology images is an instrumental step for the assessment of a tumor. Nonetheless, nucleus detection is a laborious and expensive task if done manually by experienced clinicians, and is also prone to subjectivity and inconsistency. Alternatively, the advancement in computer vision-based analysis enables the automatic detection of cancerous nuclei; however, the task poses several challenges due to the heterogeneity in the morphology and color of the nuclei, their varying chromatin distribution, and their fuzzy boundaries. In this work, we propose the usage of transformer-based detection, and dub it NucDETR, to tackle this problem, given their promising results and simple architecture on several tasks including object detection. We inspire from the recently-proposed Detection Transformer (DETR), and propose the introduction of a necessary data synthesis step; demonstrating its effectiveness and benchmarking the performance of Transformer detectors on histopathology images. Where applicable, we also propose remedies that mitigate some of the issues faced when adopting such Transformer-based detection. The proposed end-to-end architecture avoids much of the post-processing steps demanded by most current detectors, and outperforms the state-of-the-art methods on two popular datasets by 1–9% in the F-score.

Keywords: Nucleus detection · Computational histopathology · Transformer-based detection

1 Introduction

Early cancer detection and prognosis using imaging data has been major trend up to now [1,7,12,14,23]. The accurate and timely analysis of a tumor environment is critical to understand the tumor type, grading and severity, predict the survival chance of the patient, and thus explore suitable plans of treatment [21,25]. Information taken from a tumor environment is rich and heterogeneous, such as necrosis, angiogenesis, and host inflammatory response [25], which

This work is supported by research grant from ASPIRE Ref:AARE20-279.

makes the analysis of said environment highly beneficial. One such imperative assessment task is the accurate detection of nuclei in hematoxylin and eosin (H&E) stained whole-slide images (WSIs), which provides instrumental information such as the quantity of nuclei and their spatial arrangement, and can be used to facilitate and improve other downstream tasks such as high-quality segmentation of tissues [11]. Traditionally, nuclei detection is done manually, e.g. through the usage of different protein markers on the different cells in a cancerous tissue. However, such manual approaches suffer from several problems, such as low throughput, inconsistent assessment, and observer-to-observer variability [8] and require experienced personnel and extensive labor [25].

On the other hand, the continuous development of automatic, vision-based computational histology opened the door for many potentially reliable solutions. Nonetheless, automatic detection of nuclei remains challenging, and an improvement in the performance is still needed [11,15,25]. This may be caused by several problems such as poor fixation and staining or auto-focus failure during the development of the slide [25], or may occur naturally due to the complexities affected by the varying morphology or color of the nuclei, the diverse chromatin patterns, and the arrangement of nuclei in cluttered groups [11,16,25].

Examples of the tools that have been proposed to tackle the task of automatic detection of nuclei are numerous, and can be generally categorized into two categories. Firstly, earlier methods employed several classical image analysis and machine learning tools such as clustering, watershed-based segmentation, over-segmentation followed by merging, background subtraction followed by nuclear seed detection, morphological operation-based methods, support vector machines, and random forests [2,9,19,31]. The performance of earlier methods was suboptimal due to either requiring extensive development and discovery of hand-crafted features to be used with learning-based methods, or relying on a limited set of morphological structures, and failing to accommodate the versatile setting of the task.

More recently, the majority of the proposed methods employ deep learning, relying on its superior performance and automatic feature extraction. It was shown in [5] that deep learning-based methods outperform methods that rely on handcrafted features. The work in [27] employs convolutional neural networks (CNNs) aided with prior information about cell shapes, which is either supplied by an expert, or is incorporated as a trainable layer. In [4], U-net is used to extract color and shape features and thereby segment the cells in the slide image. This is followed by post-processing through erosion to refine the extracted cell boundaries. The authors of [28] also employ CNNs, but additionally utilize an unsupervised subset of the data through pseudo-labeling, thereby adopting a semi-supervised learning paradigm. In [32], LadderNet was used as a CNN base model, which is an extension of U-net, and shows an improvement in the performance. In [25], the authors propose spatially-constrained CNNs, which enforce the prediction probability of the pixels closest to the center of a nucleus. In [30], a stacked sparse auto-encoder is used as a base model, and is followed by a softmax layer to classify patches into nucleus and non-nucleus patches. The main problem with the majority of current deep learning-based detection is requiring complex pipelines that make predictions either relative to proposals, anchors, or window centers, whose design strongly affects the performance of the detector. Additionally, these pipelines often demand post processing steps to merge nearby predictions, and heuristics that assign prediction boxes to anchors, by which the performance of the proposed solution is heavily affected [3].

To address this issue, the recently proposed Detection Transformer (DETR) poses a suitable solution that combines the feature extraction prowess of CNNs, with the simplified structure of Transformers. In this paper, we propose NucDETR, a detector inspired by the DETR model, which detects nuclei in H&E WSIs, and demonstrate its superior performance to many of the current state-of-the-art methods. We aim to surpass the current performance, while simplifying the aforementioned surrogate tasks, by adopting an end-to-end pipeline that detects all nuclei in WSIs simultaneously. We prepare NucDETR by training the base CNN backbone on a simple task that classifies patches into nuclei and non-nuclei ones, prior to training it for detection. To achieve this, we propose a data synthesis routine that converts positive patches into their negative counterparts, and use this hybrid data to train the backbone. The proposed method is evaluated on two popular histology datasets and is compared against multiple methods that showed promising results on them. Our contributions in this paper can be summarized by the following:

- We propose the first Transformer-based nucleus detector in WSIs, NucDETR and benchmark its performance on two popular histopathology datasets.
- We develop a backbone training routine that familiarizes the detection pipeline with H&E stained WSIs, prior to training it for detection.
- We propose an original data augmentation method that converts positive patches into negative counterparts, thus doubling the size of the data without any labor overhead. We demonstrate the favorable effect of this step at improving the performance of the detector. This method is agnostic to the data type, and is generally suitable for data tackling cellular-level analysis, making it usable by a large body of the community.
- We compare the performance of the proposed method with multiple methods from the literature, highlight the strength and weakness points of the proposed method, and suggest several improvements where applicable, as well as future directions.

In the following, we will describe in detail the proposed methodology in Sect. 2, where we describe the data augmentation routine, provide a brief description of the detection model, and discuss some practical consideration points. In Sect. 3, we provide a description of the experimentation environment: the datasets and the compared methods, and provide pictorial and quantitative assessment of the results. Finally, we conclude with some closing remarks and suggest some potential future directions.

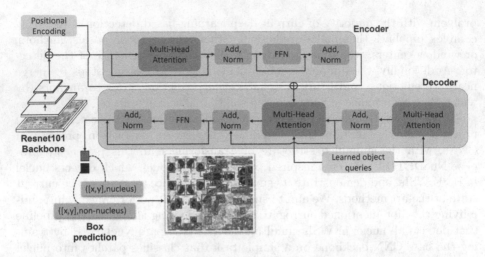

Fig. 1. NucDETR architecture, consisting of CNN model for feature extraction, positional encoding for ordinal information, object queries, and Transformer encoder-decoder modules for box-class prediction.

2   Methodology

In order to develop NucDETR for nuclei detection, two components are essential. Firstly, the base backbone which will be used to extract morphology and color features from histology slides must be familiarized with the data of interest. This is done by developing a classification version of the used data, where the backbone CNN, connected to a FFN, learns to classify each extracted patch from the WSI as either positive (consisting of nuclei) or negative. Secondly, the Transformer encoder-decoder layers must be trained on the detection data for sufficient times. The architecture in Fig. 1 summarizes the pipeline of NucDETR. As mentioned, the original DETR pipeline avoids much of the heuristics demanded by most current object detectors. This was tested empirically with the nuclei as objects-of-interest and held true to a significant degree. Nonetheless, it was also observed that including a simple one-step near-duplicate merging improves the performance by decreasing the false positive rate.

Algorithm 1 Synthesis of positive patches into negative counterparts

---

    Input: positive patches
    Output: synthetic negative patches
$R_{+ve} \leftarrow$ *set of positive regions*
for $R_i \subset R_{+ve}$ **do**
    $[w_i \times h_i] \leftarrow size(R_i)$
    $MaskNegative \leftarrow$ *extract negative mask of size* $(w_i \times h_i)$
    $R_i = MaskNegative$
end for

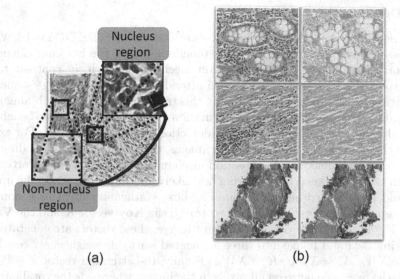

(a)   (b)

Fig. 2. Pictorial description of data synthesis (a), and sampled synthesis results from CoNSep dataset (b).

2.1 Backbone Training

To use the CNN as a feature extractor for H&E stained WSIs, it was initially trained to classify patches as either positive or negative ones. By training to do so, the CNN will learn to extract meaningful features that will be used later in the detection task. However, the main difficulty of this stage is the lack of data. As it is often the case that the vast majority of all extracted patches from a single WSI will consist of nuclei, there will be no representation of negative patches in the pool of training data. To solve this issue, we propose synthesizing negative patches out of their positive counterparts, as illustrated in Fig. 2(a). This doubles the size of the data, ensuring a balanced representation of all classes, while causing no labor overhead to obtain more data. The method used to convert positive patches into negative ones is summarized in Algorithm 1.

That is, let R_i be a positive region in a WSI with size $w_i \times h_i$, where $\bigcup_{all\ i} R_i = R_{+ve}$, and R_{+ve} is the complete set of all positive regions in a WSI. Then, in an iterative manner each R_i is replaced by a mask of the same size extracted from a neighboring negative region. Figure 2(b) demonstrates the results of the proposed synthesis method. The benefit of this approach is that the generated images are realistic and preserve the cohesion, morphology, and texture observed in the real images, ensuring that the feature extractor does not learn misleading/meaningless features. Ultimately, all nucleus incidents are eliminated; thereby, all extracted patches will only belong to the negative class. We use the binary cross-entropy loss on this hybrid data.

52   A. Obeid et al.

2.2 Detection Model

At the heart of our pipeline lies the Detection Transformer (DETR) model, which
is the second element of the detection pipeline. One of the key components in
Transformer models is the self-attention mechanism which, in contrast to its
counterpart the recurrent networks, can attend to complete and large sequences
and learn long-term relations. In images, this translates well into combining both
small and large-scale contextual information, which can be highly beneficial,
especially with the problem of nucleus detection in histology images. For example, for the object detection problem in images, the prediction of a bounding box
can be readily modeled by the attention mechanism by quantifying the attention
between two pixels, thereby answering how likely are they to represent the upper-
right and lower-left corners of a bounding box. Mathematically, this is done by
generating three vectors: the **Query** vector Q, the **Key** vector K and the **Value**
vector V [29]. Letting X denote the input image, these vectors are generated by
projecting the input image into three associated learnable weight matrices. That
is, $Q = XW_Q$, $V = XW_V$, $K = XW_K$. Finally, the attention vector $Z \in \mathbb{R}^{n \times d_v}$
models the relevance between all pixels in the image, where n is the total number
of pixels in the image, and $W_Q \in \mathbb{R}^{d \times d_q}$, $W_V \in \mathbb{R}^{d \times d_v}$, $W_K \in \mathbb{R}^{d \times d_k}$. Z is defined
as

$$Z = softmax(\frac{QK^T}{\sqrt{d_k}})V \tag{1}$$

Similar to the DETR model, we treat the task of nuclei detection in
histopathology images as a set prediction task. That is, after extracting a set
of image features, we aim to predict a *set* of bounding boxes with their class
labels in a binary classification framework, making use of the Hungarian algorithm [26], where the predicted box either consists of a nuclei, or belongs to the
null class Ø and must be discarded of. This confers the pipeline an automatic
filtering of bounding boxes and their classes to arrive at the best combination
that minimizes the detection error [3]. This is in contrast to other methods that
employ heuristics to achieve the same task [17]. It is worth mentioning that
transformer models do not depend heavily on prior information about the structure of the addressed problem as compared to their convolutional counterparts,
which well-facilitates pre-training on large corpus of unlabeled data [6].

The loss function is computed over the found optimal set as shown in Eq. (2).
$\hat{p}(c_i)$ and $\hat{b}$ denote the optimal bounding boxes and their associated classes as
found by the Hungarian algorithm. The loss function uses a linear combination of
the ℓ_1 loss and the generalized Intersection Over Union loss [24] since the latter
is scale-invariant. Had only the ℓ_1 loss been used, the scoring will be identical in
boxes that score the same overlap with the ground-truth nuclei despite differing
in scale and (hence) quality.

$$\mathcal{L}(y, \hat{y}) = \sum_{i=1}^{N} -\log(\hat{p}(c_i) + \lambda_{iou}L_{iou}(b, \hat{b}) + \lambda_{L1}\|b_i - \hat{b}_i\|, \forall i : c_i \neq \emptyset \tag{2}$$

We adopt the Transformer encoder-decoder module as shown in Fig. 1. Each encoder layer in the pipeline consists of a multi-head self-attention module followed by feed forward network (FFN), crucial to obtaining high-quality attention performance.

Practical Considerations. Upon experimentation, it was revealed that the false negative rate is small, but the false positive rate is usually large. This agrees with the observations made in [3], where the performance of DETR downgrades with small objects, i.e. the problem in hand. To that end, the following two remedies improved the performance of NucDETR. Firstly, the classification cost, that is the log-probability term in Eq. (2) has been increased, relying on the logic that since boxes that are predicted as belonging to the null class Ø are discarded of, the emphasis on this loss will lower the threshold needed to suspect a predicted box as a false alarm. Secondly, it was observed that adding a simple one-line merging of near duplicates ($\approx$1 pixel away from each other) step on the predicted boxes improves the performance of the detector. By applying both remedies, the false-alarm rate decreased, resulting in an increased precision and ultimately an increased F-score.

3 Experimental Evaluation

The proposed method NucDETR is evaluated visually and quantitatively on two popular datasets; namely, Colorectal Nuclear Segmentation and Phenotypes (CoNSep) [11], and PanNuke [10] which addresses several types of cancer. We follow the same training/testing splitting as the one suggested by the authors in both datasets. The performance of the proposed method is compared with six methods from the literature that constitute the state-of-the-art.

The used base model is Resnet-101, and is followed by [2048 $\times$ 512] and [512 $\times$ 2] fully connected layers, with Relu activation in the hidden layer, and no activation at the output layer. The model is pretrained on the COCO dataset [18], and achieves a test classification accuracy of $\sim$0.99 on the hybrid data X'.

Table 1. Quantitative assessment of proposed and state-of-the-art methods on the CoNSep and PanNuke datasets

Dataset	Method	SC-CNN [25]	CF [15]	HoVer-Net [11]	Mask-RCNN [13]	Micro-Net [22]	Dist [20]	Nuc-DETR
CoNSep	Precision	0.75	0.74	0.77	0.74	0.75	0.77	**0.80**
	Recall	0.80	0.79	0.82	0.72	0.81	0.77	**0.88**
	F1 score	0.74	0.70	0.75	0.71	0.75	0.73	**0.83**
PanNuke	Precision	0.67	–	**0.82**	0.76	0.78	0.74	0.76
	Recall	0.60	–	0.70	0.08	0.82	0.71	**0.86**
	F1 score	0.63	–	0.80	0.72	0.80	0.73	**0.81**

Similar to [25], we define True and False Positives, and True and False Negatives. Accordingly, we employ the F-score measure as the main point of

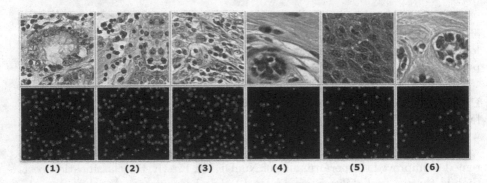

Fig. 3. Visual assessment on sampled patches form the CoNSep (the three on the left) and the PanNuke datasets. The upper row represents the original patch from the WSI. The lower row shows the ground truth nuclei in large blue circles, and the predicted ones using NucDETR in small red dots. (Color figure online)

quantitative comparison between the different methods. Additionally, we include visual assessment of the proposed method to showcase the detection of cluttered, fuzzily-bounded nuclei that also observe color variation, since such difficulties represent the main sources of challenge faced by current detectors.

3.1 Visual Assessment

Figure 3 demonstrates the prediction of NucDETR on some patches extracted from the CoNSep and PanNuke datasets. Starting off, the sampled patches belonging to CoNSep show nuclei with different boundaries embedded by tissues of different colors being successfully pointed out by NucDETR. In patch number 2, cancerous nuclei are separated from benign epithelial nuclei despite sharing similar boundary shapes. This confirms that NucDETR reliably models both morphological and textural attributes in its feature extraction. Moreover, the patches demonstrate highly cluttered nuclei being successfully detected. As for the PanNuke patches, they showcase the performance of NucDETR at detecting heavily cluttered nuclei. Both patches 4 and 6 consist of cluttered nuclei with faded and fuzzy boundaries being successfully detected. Nuclei in patch number 5 are also successfully detected noting the different shape of the nuclei and the different color of the background tissue. Interestingly, patch number 4 shows a false alarm towards the upper right corner. However, looking at the original image, there seems to be a part of a nuclei cut in half, which did not register in the ground truth of the data, but is nevertheless detected by NucDETR. Overall, the visual results of NucDETR on both datasets seem accurate.

3.2 Quantitative Assessment

We also include a quantitative comparison between the proposed and the state-of-the-art methods in Table 1. Starting with the CoNSep dataset, the proposed method outperforms the second best method i.e. HoVer-Net [11] by a factor of 9% in the F-score. The precision score is also higher by a considerable factor; the recall score is also higher, but to a lesser degree. A similar trend is observed in PanNuke, where the F-score of the proposed method is slightly higher than the state-of-the-art performance in HoVer-Net. Nevertheless, there is ~9% increase in the recall above HoVer-Net, accompanied with a drop in the precision, alluding to the discussed issue observed with NucDETR, where a small false negative rate is achieved albeit with a larger false positive rate. By benchmarking the performance of NucDETR, we hope to highlight its shortcomings and strengths, which may open the door for further improvements in the future.

All in all, the increased F-score observed in both tables is enough to conclude that the proposed transformer-based detector NucDETR is more robust to color and morphology variations, the heteregenous chromatin patterns, and is able to better model and detect fuzzy and irregular boundaries, and is more robust to cluttered nuclei. Overall, this ensures a more accurate detection of the nuclei in histopathology images, and consequently a more reliable performance in downstream steps, such as tissue segmentation, patient survival prediction, or exploration of treatment plans.

Conclusion

In this work, we proposed NucDETR, the first Transformer-based nuclei detector in histopathology images, and benchamrked its performance on two datasets. We developed a data augmentation routine that helped familiarize NucDETR with histopathology data prior to training it for detection. We observed the promising performance of the proposed model, as well as its potential shortcomings. As for the latter, we proposed the usage of a simple merging step that increased the precision of the model. Both quantitative and visual evaluations confirm the validity of the proposed model, and suggest that it is highly competitive with the state-of-the-art, clearly outperforming existing methods in the F-score. At last, we believe that there are several subtopics to be addressed that may have a direct positive impact on the performance of Transformer detectors at nucleus detection. Namely, it may be beneficial to investigate other methods to increase the size of the data used to train the backbone CNN feature extractor, beyond the suggested method that has an upper limit in the size of the hybrid data of $2 * n$. An example would be the usage of GAN-based data synthesis. Moreover, other variants of the DETR model may perform better with smaller training schedules such as Deformable DETR. Lastly, we beleive that pretraining NucDETR on unlabeled data may further improve its performance. We hope that the presented work and the suggested future directions would open the door for further improvements in this field.

References

1. Alkadi, R., Taher, F., El-baz, A., Werghi, N.: A deep learning-based approach for the detection and localization of prostate cancer in T2 magnetic resonance

images. J. Digit. Imaging **32**(5), 793–807 (2018). https://doi.org/10.1007/s10278-018-0160-1

2. Bell, A.A., Herberich, G., Meyer-Ebrecht, D., Bocking, A., Aach, T.: Segmentation and detection of nuclei in silver stained cell specimens for early cancer diagnosis. In: 2007 IEEE International Conference on Image Processing, vol. 6, pp. VI–49. IEEE (2007)

3. Carion, N., Massa, F., Synnaeve, G., Usunier, N., Kirillov, A., Zagoruyko, S.: End-to-end object detection with transformers. In: Vedaldi, A., Bischof, H., Brox, T., Frahm, J.-M. (eds.) ECCV 2020. LNCS, vol. 12346, pp. 213–229. Springer, Cham (2020). https://doi.org/10.1007/978-3-030-58452-8_13

4. Chen, K., Zhang, N., Powers, L., Roveda, J.: Cell nuclei detection and segmentation for computational pathology using deep learning. In: 2019 Spring Simulation Conference (SpringSim), pp. 1–6. IEEE (2019)

5. Cruz-Roa, A.A., Arevalo Ovalle, J.E., Madabhushi, A., González Osorio, F.A.: A deep learning architecture for image representation, visual interpretability and automated basal-cell carcinoma cancer detection. In: Mori, K., Sakuma, I., Sato, Y., Barillot, C., Navab, N. (eds.) MICCAI 2013. LNCS, vol. 8150, pp. 403–410. Springer, Heidelberg (2013). https://doi.org/10.1007/978-3-642-40763-5_50

6. Dai, Z., Cai, B., Lin, Y., Chen, J.: UP-DETR: unsupervised pre-training for object detection with transformers. In: Proceedings of the IEEE/CVF Conference on Computer Vision and Pattern Recognition, pp. 1601–1610 (2021)

7. ElKhatib, O., Werghi, N., Al-Ahmad, H.: Automatic polyp detection: a comparative study. In: Annual International Conference of the IEEE Engineering in Medicine and Biology Society, pp. 2669–2672 (2019)

8. Elmore, J.G., et al.: Diagnostic concordance among pathologists interpreting breast biopsy specimens. JAMA **313**(11), 1122–1132 (2015)

9. Filipczuk, P., Kowal, M., Obuchowicz, A.: Automatic breast cancer diagnosis based on k-means clustering and adaptive thresholding hybrid segmentation. In: Choras, R.S. (eds) Image Processing and Communications Challenges 3. Advances in Intelligent and Soft Computing, vol. 102. Springer, Heidelberg (2011). https://doi.org/10.1007/978-3-642-23154-4_33

10. Gamper, J., et al.: Pannuke dataset extension, insights and baselines. arXiv preprint arXiv:2003.10778 (2020)

11. Graham, S., et al.: Hover-Net: simultaneous segmentation and classification of nuclei in multi-tissue histology images. Med. Image Anal. **58**, 101563 (2019)

12. Hassan, T., et al.: Nucleus classification in histology images using message passing network. Med. Image Anal. **79**, 102480 (2022)

13. He, K., Gkioxari, G., Dollár, P., Girshick, R.: Mask R-CNN. In: Proceedings of the IEEE International Conference on Computer Vision, pp. 2961–2969 (2017)

14. Javed, S., Mahmood, A., Dias, J., Werghi, N.: Multi-level feature fusion for nucleus detection in histology images using correlation filters. Comput. Biol. Med. **143**, 105281 (2022)

15. Javed, S., Mahmood, A., Dias, J., Werghi, N., Rajpoot, N.: Spatially constrained context-aware hierarchical deep correlation filters for nucleus detection in histology images. Med. Image Anal. **72**, 102104 (2021)

16. Jung, H., Lodhi, B., Kang, J.: An automatic nuclei segmentation method based on deep convolutional neural networks for histopathology images. BMC Biomed. Eng. **1**(1), 1–12 (2019)

17. Khan, S., Naseer, M., Hayat, M., Zamir, S.W., Khan, F.S., Shah, M.: Transformers in vision: a survey. ACM Comput. Surv. (2021)

18. Lin, T.Y., et al.: Microsoft COCO: Common Objects in Context. In: Fleet, D., Pajdla, T., Schiele, B., Tuytelaars, T. (eds.) ECCV 2014. LNCS, vol. 8693, pp. 740–755. Springer, Cham (2014). https://doi.org/10.1007/978-3-319-10602-1_48
19. Mao, K.Z., Zhao, P., Tan, P.H.: Supervised learning-based cell image segmentation for P53 immunohistochemistry. IEEE Trans. Biomed. Eng. **53**(6), 1153–1163 (2006)
20. Naylor, P., Laé, M., Reyal, F., Walter, T.: Segmentation of nuclei in histopathology images by deep regression of the distance map. IEEE Trans. Med. Imaging **38**(2), 448–459 (2018)
21. O'Brien, C.A., Pollett, A., Gallinger, S., Dick, J.E.: A human colon cancer cell capable of initiating tumour growth in immunodeficient mice. Nature **445**(7123), 106–110 (2007)
22. Raza, S.E.A., et al.: Micro-Net: a unified model for segmentation of various objects in microscopy images. Med. Image Anal. **52**, 160–173 (2019)
23. Reda, I., et al.: Computer-aided diagnostic tool for early detection of prostate cancer. In: IEEE International Conference on Image Processing. IEEE (2016)
24. Rezatofighi, H., Tsoi, N., Gwak, J., Sadeghian, A., Reid, I., Savarese, S.: Generalized intersection over union: a metric and a loss for bounding box regression. In: Proceedings of the IEEE/CVF Conference on Computer Vision and Pattern Recognition, pp. 658–666 (2019)
25. Sirinukunwattana, K., Raza, S.E.A., Tsang, Y.W., Snead, D.R., Cree, I.A., Rajpoot, N.M.: Locality sensitive deep learning for detection and classification of nuclei in routine colon cancer histology images. IEEE Trans. Med. Imaging **35**(5), 1196–1206 (2016)
26. Stewart, R., Andriluka, M., Ng, A.Y.: End-to-end people detection in crowded scenes. In: Proceedings of the IEEE Conference on Computer Vision and Pattern Recognition, pp. 2325–2333 (2016)
27. Tofighi, M., Guo, T., Vanamala, J.K., Monga, V.: Prior information guided regularized deep learning for cell nucleus detection. IEEE Trans. Med. Imaging **38**(9), 2047–2058 (2019)
28. Valkonen, M., Högnäs, G., Bova, G.S., Ruusuvuori, P.: Generalized fixation invariant nuclei detection through domain adaptation based deep learning. IEEE J. Biomed. Health Inform. **25**(5), 1747–1757 (2020)
29. Vaswani, A., et al.: Attention is all you need. Adv. Neural Inf. Process. Syst. **30** (2017)
30. Xu, J., Xiang, L., Hang, R., Wu, J.: Stacked sparse autoencoder (SSAE) based framework for nuclei patch classification on breast cancer histopathology. In: 2014 IEEE 11th International Symposium on Biomedical Imaging (ISBI), pp. 999–1002. IEEE (2014)
31. Yang, X., Li, H., Zhou, X.: Nuclei segmentation using marker-controlled watershed, tracking using mean-shift, and Kalman filter in time-lapse microscopy. IEEE Trans. Circuits Syst. I Regul. Pap. **53**(11), 2405–2414 (2006)
32. Zhuang, J.: LadderNet: multi-path networks based on U-Net for medical image segmentation. arXiv preprint arXiv:1810.07810 (2018)

Self-supervised Learning Based on a Pre-trained Method for the Subtype Classification of Spinal Tumors

Menglei Jiao[1,6], Hong Liu[1]([✉]), Zekang Yang[1,6], Shuai Tian[2], Hanqiang Ouyang[3,4,5], Yuan Li[2], Yuan Yuan[2], Jianfang Liu[2], Chunjie Wang[2], Ning Lang[2], Liang Jiang[3,4,5], Huishu Yuan[2], Yueliang Qian[1], and Xiangdong Wang[1]

[1] Beijing Key Laboratory of Mobile Computing and Pervasive Device, Institute of Computing Technology, Chinese Academy of Sciences, Beijing 100190, China
hliu@ict.ac.cn
[2] Department of Radiology, Peking University Third Hospital, Beijing 100191, China
[3] Department of Orthopaedics, Peking University Third Hospital, Beijing 100191, China
[4] Engineering Research Center of Bone and Joint Precision Medicine, Beijing 100191, China
[5] Beijing Key Laboratory of Spinal Disease Research, Beijing 100191, China
[6] University of Chinese Academy of Sciences, Beijing 100086, China

Abstract. Spinal tumors contain multiple pathological subtypes, and different subtypes may correspond to different treatments and prognoses. Diagnosis of spinal tumor subtypes from medical images in the early stage is of great clinical significance. Due to the complex morphology and high heterogeneity of spinal tumors, it can be challenging to diagnose subtypes from medical images accurately. In recent years, a number of researchers have applied deep learning technology to medical image analysis. However, such research usually requires a large number of labeled samples for training, which can be difficult to obtain. Therefore, the use of unlabeled medical images to improve the identification performance of models is a hot research topic. This study proposed a self-supervised learning based pre-trained method Res-MAE using a convolutional neural network and masked autoencoder. First, this method trains an efficient feature encoder using a large amount of unlabeled spinal medical data with an image reconstruction task. Then this encoder is transferred to the downstream subtype classification in a multi-modal fusion model for fine-tuning. This multi-modal fusion model adopts a bipartite graph and multi-branch for spinal tumor subtype classification. The experimental results show that the accuracy of the proposed method can be increased by up to 10.3%, and the F1 can be increased by up to 13.8% compared with the baseline method.

Keywords: Self-supervised learning · Pre-training model · Subtype classification

1 Introduction

Spinal tumors occur in the spine and include primary and metastatic spinal tumors. There are dozens of subtypes of primary spinal tumors, including giant cell tumors of

W. Qin et al. (Eds.): CMMCA 2022, LNCS 13574, pp. 58–67, 2022.
https://doi.org/10.1007/978-3-031-17266-3_6

bone, myeloma, and chordoma [1]. Different tumor subtypes can cause different damage and correspond to different treatments. Diagnosis of spinal tumor subtypes from medical images in the early stage is of great clinical significance. Due to the complex morphology and high heterogeneity of spinal tumors, it can be challenging to diagnose subtypes from medical images accurately. With the development of deep learning technology, a number of researchers have applied natural image technology to medical images. Although it is difficult to obtain labeled medical images, there are a large number of unlabeled medical images. Many studies have been conducted on how to use this unlabeled data to train models to improve classification performance.

Using a limited number of labeled medical images and a large number of unlabeled medical images to classify tumors can be problematic with a small sample. Currently, the more common methods used are semi-supervised learning [2] and self-supervised learning [3]. Semi-supervised learning means that when training the model, not only labeled data can be used, but also unlabeled data can be used. Liu et al. [4] proposed a relationship driven semi-supervised medical image classification framework based on a consistency method. Unlabeled data can be used by encouraging the prediction consistency of a given input under disturbance. Furthermore, the built-in disturbance model is used to generate high-quality consistency targets for unlabeled data. Tseng et al. [5] proposed a new convolutional neural network DNetUNet, which combines U-Net [6] with different down-sampling levels and a new dense block as a feature extractor. The generated adversarial network generates pseudo labels for a large number of unlabeled data and adds them to the training of the model. The experimental results show that after applying a large number of unlabeled data, the performance can be improved.

Self-supervised learning refers to the use of unlabeled data for model training, which is usually used to build a pre-trained model. Based on the visual transformer (ViT) [8], He et al. proposed an unsupervised learning method based on masked autoencoder (MAE) [10]. Through the image reconstruction task, some patches are randomly masked during training, and only the features of the remaining patches are extracted. The designed decoder can use the features extracted by the encoder to restore the image. To make the recovery effect good enough, the encoder needs to be able to extract the key features as much as possible [10]. This method constructs an efficient feature encoder through unsupervised learning on ImageNet [11] and transfers the encoder to detection, segmentation, classification and other tasks to achieve the best results.

For medical image analysis, the multi-modal fusion strategy can also improve the performance of the model; for example, the fusion of axial magnetic resonance imaging (MRI) and ultrasound for volume registration [12], fusion of axial MRI and positron emission computed tomography (CT) for automated dementia diagnosis [13], and fusion of different sequences of axial CT and axial MRI for tumor segmentation [14–16]. However, the methods mentioned above focus on the fusion of single planes and do not comprehensively consider the association between different planes.

Due to the complex morphology and high heterogeneity of spinal tumors, it is challenging to accurately classify subtypes from medical images, especially in a small sample. In this study, a self-supervised learning method for medical images is proposed to construct an efficient feature encoder and transfer the feature encoder to the downstream multi-modal spinal tumor subtype classification task. The effectiveness of the proposed method was verified using a multi-modal spinal tumor subtype dataset.

2   Proposed Method

This report will introduce the pre-trained method based on self-supervised learning and the multi-modal fusion model.

2.1   Pre-trained Method Based on Self-supervised Learning

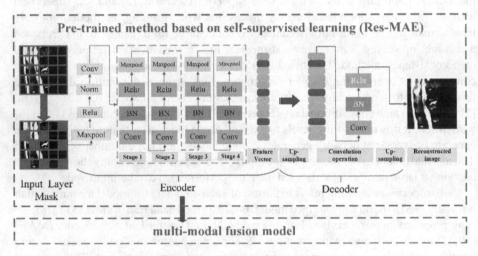

Fig. 1. Framework of Res-MAE

It can increase the sample scale to improve the classification performance when the sample scale of the dataset (target domain) is small. Alternatively, a pre-trained model can be built on a dataset (source domain) with a large number of samples, after which the pre-trained model is transferred to the target domain [7]. Recently, an increasing number of pre-training methods have adopted visual attention models [8, 9]. The MAE method [10] applied the image reconstruction task to get the efficient feature encoder and transfer it to downstream tasks. However, MAE is completely based on the attention mechanism [9], which requires a large number of samples to train a pre-trained model and cannot adapt to a small sample of medical images. This study proposes a pre-trained method based on a convolution neural network and MAE (Res-MAE) combined with multi-modal fusion for spinal tumor subtype classification.

The overall framework of Res-MAE is shown in Fig. 1, which contains an input layer mask, encoder, decoder, and loss function. The input layer is used to randomly mask some patches in the spinal medical images during training. The encoder is used to extract the key features using a Res-MAE. The decoder is used to reconstruct the image, and the loss function is used to supervise the training process. The trained encoder will be transferred to the convolutional neural network branch in the following multi-modal fusion model for the spinal tumor subtype classification task.

Input Layer Mask: In the input layer mask model, before the image is sent to the deep learning model, the image is first cut into several patches with the size of $h \times w$. Then, a certain proportion of α is used to mask some patches, as shown in Fig. 1. The purpose is to enable the decoder to recover the original image through the extracted features as much as possible when the image information is partially missing. Assume that the size of the input image is $H \times W$. To realize fast image cutting, the image is directly matrix transformed. Let $n = \frac{H}{h} \times \frac{W}{w}$, which represents the number of patches. It will get a matrix with the shape of $n \times h \times w$, which represents that the raw image is cut into n patches with the size of $h \times w$. Then, these patches should be expanded in one dimension. The next step is the mask operation. Specifically, several patches should be set as 0, and the specific number of masks is determined by the scale coefficient α to control. Finally, after the mask operation is completed, the matrix is restored, and the next encoding operation is performed. Generally speaking, the better the image quality restored by the decoder, the more representative the features extracted by the encoder are.

Encoder and Decoder: The structure of the encoder and decoder is shown in Fig. 1. The encoder used in this part is ResNet18, which has four stages. Each stage includes several convolution operations (Conv), standardization operations (BN), activation function operations (Relu), and maximum pool operations (Maxpool). The maximum pool operation is used to select the feature with the largest response value as the input for the next stage. Thus, each stage can extract features at different scales. In this study, the feature vector output in the last stage of ResNet18 is used as the input of the decoder. To prevent the decoder from having strong memory of specific features and causing overfitting, the decoder structure used in this study is very simple, that is, up-sampling, convolution, and up-sampling. The purpose of two up-sampling is to make the output size of the decoder equal to the size of the original image. The purpose of convolution operation is to add learnable parameters to the decoder. Finally, the decoder outputs the reconstructed image with the same size as the original image and calculates the reconstruction loss with the original image. The reconstruction loss function is introduced below.

Loss Function: The reconstructed image and the original image can be compared pixel by pixel to measure the quality of reconstruction. The closer the pixel value at the corresponding position is, the better the reconstruction effect is. Based on this, this study uses the mean square error (MSE) loss as the reconstruction loss between the reconstructed image and the original image, as shown in Eq. 1, where n represents the number of pixels, $\hat{y}$ represents the reconstructed image, and y represents the original image. The better the reconstruction effect, the smaller the L value. The goal of optimization is to make L as small as possible.

$$L = \frac{1}{n} \sum\nolimits_{i=1}^{n} \left(\hat{y}_i - y_i \right)^2 \qquad (1)$$

2.2 Multi-modal Fusion Model

Patients often have image data from different modalities, such as MRI and CT, and different planes, including axial and sagittal. Different modalities may contain abundant information on tumors, which can improve the deep model to classify tumor subtypes. This study uses a multi-modal fusion model based on the pre-trained model and fuses the multi-modal image data in the input, feature, and decision-making layers, respectively. As shown in Fig. 2, in the input layer, the image data from two different modalities of patients are constructed into a bipartite graph structure, and the images of different modalities connected by each edge are used as the input of the feature layer. The features of different modalities are extracted and fused through the convolution neural network branch and vision transformer branch. Finally, the predicted values of each patient are fused at the decision-making level through trusted edge set filtering to get the tumor subtype at the patient level.

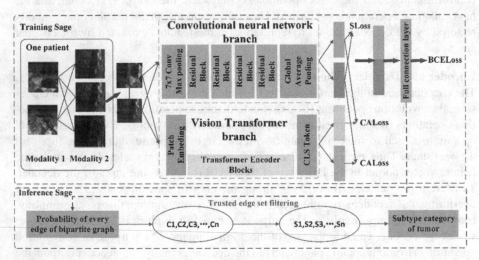

Fig. 2. Multi-modal fusion model

In the training stage, as shown in Eq. 2 and Eq. 3, *CALoss* is used to make the features extracted by the two branches for the same modality as similar as possible, where $C_{i,j}$ and $A_{i,j}$ represent the output of the convolutional neural network branch and vision transformer branch, respectively, and $CALoss_1$ and $CALoss_2$ will be obtained respectively. *SLoss* can make different modal features of the same patients closer and different patients farther, where A represents the type of center of first modality features of patients in each batch and S represents the type of center of second modality features. *BCELoss* is used to calculate the final classification loss. Finally, the loss function of the whole deep learning model is shown in Eq. 4, where a, b and c are the weights of each part.

$$CALoss = \frac{\sqrt{\sum_j \left(C_{i,j} - A_{i,j} \right)^2}}{d} \tag{2}$$

$$SLoss = \left\| AS^T - I \right\|_2 \tag{3}$$

$$Loss = \alpha \cdot (CALoss_1 + CALoss_2) + b \cdot SLoss + c \cdot BCELoss \tag{4}$$

2.3 Pre-trained Model Transfer to the Multi-modal Fusion Model

According to the pre-trained method Res-MAE proposed in Sect. 2.1, the convolution neural network is used as the encoder to construct the pre-trained model on a large number of spinal medical image datasets, and the MSE is used as the image reconstruction loss. After training, this encoder can extract the key features of spinal medical images. Then, the weight of the encoder is transferred to the multi-modal fusion model for spinal tumor subtype classification. The specific transfer method is used to save the trained encoder weight and restore the saved pre-trained weight to the convolution neural network branch in the multi-modal fusion model. Then on the limited labeled spinal tumor subtype dataset, this encoder is further trained, and the weight parameters of the model are adjusted to improve the classification performance for the spinal tumor subtype task.

3 Experiment and Results

Table 1. Subtype information of D-Dataset

Tumor types	Training set	Test set
Schwannoma/neurofibroma	82	41
Myeloma	38	19
Giant cell tumor of bone	34	16
Chordoma	24	12
Langerhans	19	9
Total	**197**	**97**

Dataset: This study used the P-Dataset to construct the pre-trained model, including the image data of 962 cases from a hospital. This dataset contained a total of 63,991 images, including metastatic and primary spinal tumors with two different scanning modalities of CT and MRI and three different scanning planes (axial, sagittal, and coronal). In addition, the experiment on the spinal tumor subtype classification task was conducted using D-Dataset, which contains five common spinal tumor subtypes. All cases in D-Dataset contained axial and sagittal MRI, and the case number of each subtype for the training and test is shown in Table 1.

Experimental Design: First, the self-supervised pre-trained model is constructed based on the P-Dataset, and then the obtained encoder weights are transferred to the multi-modal fusion model. The classification experiments of spinal tumors are conducted on the D-Dataset. We conducted comparative experiments with and without a pre-trained model on D-Dataset. At the same time, the corresponding experiments were conducted in different h, w, α to verify the effectiveness of the method, where h and w respectively represent the height and width of the patches in the process of image reconstruction and α represents the proportion of patches masked. The self-supervised pre-trained model used a learning rate of 0.2 in the training stage and stochastic gradient descent (SGD) as the optimizer to train 100 epochs. When the trained encoder weights are transferred to the spinal tumor subtype classification task, the SGD is used as the optimizer to train 20 epochs.

Metrics: The common accuracy (ACC), F1, precision (Pre), and recall (Rec) of the multi-classification model are used as the evaluation metrics.

Experimental Results Under Different Methods: Table 2 shows the experimental results for the classification of five spinal tumor subtypes on D-Dataset with and without the Res-MAE pre-trained method and with a different selection of h, w, α in Res-MAE. It can be seen from Table 2 that after using Res-MAE to build the pre-trained model and transfer the encoder to the downstream subtype task, the ACC and F1 values improved by about 3.1%~10.3% and about 6.8%~13.8% and Pre and Rec by about 0.9%~7.9%, and about 5.6%~15.5%. It can be seen from Table 2 that when the mask coverage is 75% and $h = w = 1$, the model performs best. Compared with the model without pre-training, ACC increases from 56.7% to 67.0%, and F1 from 41.6% to 55.4%. Compared with multi-modal fusion, in the case of $\alpha = 0.75$, $h = w = 1$, the ACC using only axial plane images is 58.8%, and the ACC using only sagittal plane images is 60.8%, both lower than that of multi-modal fusion. It can be seen from the experimental results that the performance of the model classification can be improved by building the pre-trained model through Res-MAE and transferring it to the spinal tumor subtype classification task.

Comparison Between Models and Doctors: To verify the effectiveness of the model, we invited three doctors (D1, D2, and D3 with three, 12, and eight years of experience, respectively) to identify the tumor subtypes of patients on the test set according to the images. The results are shown in Table 3.

Without using the Res-MAE pre-trained model, the ACC of the multi-modal fusion model was 3.1% higher than D1, 6.2% higher than D3, and lower than D2, and the F1 was 3.4% higher than D3 and lower than D1 and D2. After using the Res-MAE pre-trained model, the multi-modal fusion model was higher than all the three doctors in all metrics. The ACC and F1 values were up to 16.5% and 17.2% higher, Pre and Rec up to 20.1% and 17.3% higher. These results show the effectiveness of our proposed pre-trained based multi-modal fusion method.

To analyze the differences between the doctors and model in more detail, we counted and drew the confusion matrix between the doctors and model. As shown in Fig. 3, both the doctors and model had the highest classification accuracy in relation to schwannoma/neurofibroma, which may be due to this subtype having significant visual features

Table 2. Experimental results on D-Dataset

	Methods	ACC(%)	F1(%)	Pre(%)	Rec(%)
Axial & sagittal	No pre-trained model	56.7	41.6	53.9	40.6
	$\alpha = 0.25, h = w = 8$	62.9	50.7	54.8	50.9
	$\alpha = 0.50, h = w = 8$	65.0	52.3	56.1	52.1
	$\alpha = 0.75, h = w = 8$	62.9	50.6	55.8	50.2
	$\alpha = 0.85, h = w = 8$	59.8	48.4	58.5	46.2
	$\alpha = 0.75, h = w = 1$	**67.0**	**55.4**	**61.8**	**55.7**
	$\alpha = 0.75, h = w = 4$	66.0	55.1	59.6	56.1
	$\alpha = 0.75, h = w = 16$	65.0	52.9	55.1	54.2
	$\alpha = 0.75, h = w = 32$	63.9	51.9	53.1	54.7
Axial	$\alpha = 0.75, h = w = 1$	58.8	46.9	47.2	47.9
Sagittal	$\alpha = 0.75, h = w = 1$	60.8	55.8	60.6	54.1

Table 3. Comparison results between the model and doctors on D-Dataset

Doctors/Model	ACC	F1	Pre	Rec
D1	53.6	43.2	52.4	43.2
D2	66.0	51.8	59.0	50.9
D3	50.5	38.2	41.7	38.4
Multi-modal fusion model	56.7	41.6	53.9	40.6
Multi-modal fusion model with Res-MAE	**67.0**	**55.4**	**61.8**	**55.7**

and a higher prevalence. In addition, the classification accuracy of the model on myeloma was higher than that of the three doctors, and the classification accuracy of giant cell tumors of bone, chordoma, and langerhans was close to that of the doctors. The main reason for this phenomenon may be that the number of samples of the latter three sub-types was too small, and the model could not learn a sufficient number of distinguishing features. However, the model is still close to doctors in the classification of these tumor types, indicating that the model can reach a level close to doctors through the learning of limited samples.

4 Conclusion

This study proposes a pre-trained method Res-MAE based on a convolutional neural network and MAE, which constructs an efficient feature encoder on a large number of unlabeled medical image data through a self-supervised learning and image reconstruction task. The encoder is transferred to the multi-modal fusion model of the downstream

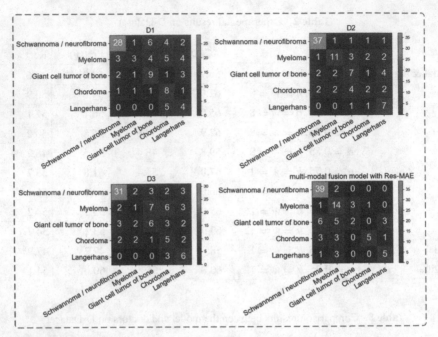

Fig. 3. Confusion matrix of the doctors and model for five subtypes

spinal tumor subtype classification task for transfer learning. The experimental results show that the feature encoder constructed by Res-MAE can improve the performance of downstream classification tasks to a certain extent. In addition, the proposed method has certain advantages compared to doctors.

Acknowledgement. This work was supported by the Beijing Natural Science Foundation (Z190020), National Natural Science Foundation of China (82171927, 81971578), Capital's Funds for Health Improvement and Research (2020-4-40916).

References

1. Shrivastava, R.K., Epstein, F.J., Perin, N.I., et al.: Intramedullary spinal cord tumors in patients older than 50 years of age: management and outcome analysis. J. Neurosurg. Spine **2**(3), 249–255 (2005)
2. Huang, Z., et al.: Universal semi-supervised learning. Adv. Neural Inf. Process. Syst. **34**, 26714–26725 (2021)
3. Chowdhury, A., et al.: Applying self-supervised learning to medicine: review of the state of the art and medical implementations. In: Informatics, vol. 8, no. 3. MDPI (2021)
4. Liu, Q., Yu, L., Luo, L., et al.: Semi-supervised medical image classification with relation-driven self-ensembling model. IEEE Trans. Med. Imaging **39**(11), 3429–3440 (2020)
5. Tseng, K.K., Zhang, R., Chen, C.M., et al.: DNetUNet: a semi-supervised CNN of medical image segmentation for super-computing AI service. J. Supercomput. **77**(4), 3594–3615 (2021)

6. Ronneberger, O., Fischer, P., Brox, T.: U-net: convolutional networks for biomedical image segmentation. In: Navab, N., Hornegger, J., Wells, W.M., Frangi, A.F. (eds.) MICCAI 2015. LNCS, vol. 9351, pp. 234–241. Springer, Cham (2015). https://doi.org/10.1007/978-3-319-24574-4_28

7. Valverde, J.M., et al.: Transfer learning in magnetic resonance brain imaging: a systematic review. J. Imaging 7(4), 66 (2021)

8. Dosovitskiy, A., Beyer, L., Kolesnikov, A., et al.: An image is worth 16x16 words: transformers for image recognition at scale. arXiv preprint arXiv:2010.11929 (2020)

9. Vaswani, A., et al.: Attention is all you need. Adv. Neural Inf. Process. Syst. **30**, 6000–6010 (2017)

10. He, K., Chen, X., Xie, S., et al.: Masked autoencoders are scalable vision learners. arXiv preprint arXiv:2111.06377 (2021)

11. Russakovsky, O., et al.: Imagenet large scale visual recognition challenge. Int. J. Comput. Vis. **115**(3), 211–252 (2015)

12. Song, X., et al.: Cross-modal attention for mri and ultrasound volume registration. In: de Bruijne, M., et al. (eds.) MICCAI 2021. LNCS, vol. 12904, pp. 66–75. Springer, Cham (2021). https://doi.org/10.1007/978-3-030-87202-1_7

13. Zhou, T., et al.: Deep multi-modal latent representation learning for automated dementia diagnosis. In: Shen, D., et al. (eds.) MICCAI 2019. LNCS, vol. 11767, pp. 629–638. Springer, Cham (2019). https://doi.org/10.1007/978-3-030-32251-9_69

14. Zhang, Y., et al.: Modality-aware mutual learning for multi-modal medical image segmentation. In: de Bruijne, M., et al. (eds.) MICCAI 2021. LNCS, vol. 12901, pp. 589–599. Springer, Cham (2021). https://doi.org/10.1007/978-3-030-87193-2_56

15. Zhang, Y., et al.: Multi-phase liver tumor segmentation with spatial aggregation and uncertain region inpainting. In: de Bruijne, M., et al. (eds.) MICCAI 2021. LNCS, vol. 12901, pp. 68–77. Springer, Cham (2021). https://doi.org/10.1007/978-3-030-87193-2_7

16. Syazwany, N.S., Nam, J.-H., Lee, S.-C.: MM-BiFPN: multi-modality fusion network with Bi-FPN for MRI brain tumor segmentation. IEEE Access **9**, 160708–160720 (2021)

CanDLE: Illuminating Biases in Transcriptomic Pan-Cancer Diagnosis

Gabriel Mejía[(✉)], Natasha Bloch, and Pablo Arbelaez

Center for Research and Formation in Artificial Intelligence, Universidad de los Andes, Bogotá, Colombia
{gm.mejia,n.blochm,pa.arbleaez}@uniandes.edu.co

Abstract. Automatic cancer diagnosis based on RNA-Seq profiles is at the intersection of transcriptome analysis and machine learning. Methods developed for this task could be a valuable support in clinical practice and provide insights into the cancer causal mechanisms. To correctly approach this problem, the largest existing resource (The Cancer Genome Atlas) must be complemented with healthy tissue samples from the Genotype-Tissue Expression project. In this work, we empirically prove that previous approaches to joining these databases suffer from translation biases and correct them using batch z-score normalization. Moreover, we propose CanDLE, a multinomial logistic regression model that achieves state of the art performance in multilabel cancer/healthy tissue type classification (94.1% balanced accuracy) and all-vs-one cancer type detection (78.0% average max F_1).

Keywords: Cancer classification · Cancer detection · Machine learning · Multinomial logistic regression · TCGA · GTEx

1 Introduction

Over the last decade, the fast advances in genome sequencing technologies have proven revolutionary, promoting improvements in experimental and high throughput techniques related to transcriptome analysis and bioinformatics [17]. This effort has led to an increase in public RNA-Seq data available to the cancer research community [2]. Consequently, large datasets like The Cancer Genome Atlas (TCGA) [1] have been established, serving as a valuable framework for obtaining standardized and curated genetic expression profiles. This data abundance, combined with the recent success of machine learning in medical applications, makes the automatic cancer diagnosis from transcriptomic samples more plausible than ever before.

There have been numerous approaches to this problem [2,3,10,12,13,15] using a wide range of techniques that go from classic machine learning algorithms (e.g., K-nearest neighbors [10]) to cutting edge deep learning models (e.g., graph convolutional neural networks [15]). However, most of these works are trained and tested exclusively in the TCGA, which has an extremely low

© The Author(s), under exclusive license to Springer Nature Switzerland AG 2022
W. Qin et al. (Eds.): CMMCA 2022, LNCS 13574, pp. 68–77, 2022.
https://doi.org/10.1007/978-3-031-17266-3_7

number of healthy tissue samples ($\approx 7\%$). Due to this limitation, almost all methods aggregate healthy samples from different tissues in a single class and solve a multilabel classification of 34 classes (33 cancer types and 1 small healthy class). These experimental conditions are far from what is observed in clinical practice and, consequently, lower the applicability of the results. Moreover, given the significant differences of gene expression profiles between tissues, a tissue classifier will perform adequately in this framework without learning to discriminate between cancer and healthy tissue.

One way to improve the practical usability of the models is to include a comparable and paired amount of healthy samples. For this task, the Genotype-Tissue Expression project (GTEx) is the natural choice since it has a similar scale to the TCGA and captures RNA-Seq samples from non-diseased tissue sites [11]. Aiming to provide a common database, recent works [18,19] apply standardized quantification and normalization to raw data and obtain joint TCGA and GTEx cohorts. Using these resources, novel research has been published performing cancer detection [14], or multi-task cancer type classification [7] and achieving outstanding results.

Although these advances are clear steps in the right direction, two main problems need to be addressed in the current state of the art: (1) there is no empirical proof of the absence of translation biases in the joint datasets, and (2) there is no clear evidence of an important metric improvement associated to the use of cutting edge deep learning techniques compared to simple algorithms. Of these, the first issue is of cornerstone importance because if the origin sources are linearly separable, then the problem would again be in its over-simplified form where it is enough to separate the GTEx and TCGA and then perform tissue classification to obtain great results.

To address both problems, in this work, we first empirically prove the existence of such bias and correct it using batch z-score normalization, which is widely adopted by the machine learning community [16]. And secondly, we propose a simple multinomial logistic regression method to perform multilabel cancer/healthy tissue type classification with sound performance. Our model, which we call CanDLE (Cancer Diagnosis Logistic Engine), cannot only be used for multilabel classification but also highly unbalanced specific cancer type detection.

Our main contributions can be summarized as follows:

1. We empirically prove that previous approaches to joining the GTEx and TCGA databases suffer from significant biases and use a simple batch normalization technique to correct them.
2. We show that a simple method such as multinomial logistic regression can obtain state of the art performance (94.1% balanced accuracy) in multilabel classification of cancerous and healthy tissue types.
3. We demonstrate that CanDLE can detect specific cancer types in highly unbalanced scenarios with state of the art performance (78.0% average max F_1).

4. We exploit the simplicity of our method to perform intuitive and direct gene relevance interpretation in pan-cancer classification.

To ensure the reproducibility of our results, all the resources of this paper are publicly available in https://github.com/g27182818/CanDLE.

2   Related Work

2.1   Joining the TCGA and GTEx Databases

Both the TCGA and GTEx are the gold standard databases for publicly available RNA-Seq profiles of cancerous and healthy tissue, respectively. The TCGA has processed more than $10,000$ samples spanning 33 cancer types, and healthy tissue controls [1], and the GTEx project has collected samples from 54 non-diseased tissue sites across nearly $1,000$ individuals [11]. However, differences in alignment, quantification, and normalization protocols had prevented the use of both databases in joint transcriptomic analyses.

Vivian et al. [18] were the first to propose a unified GTEx-TCGA dataset. They performed standardized alignment with STAR [4] and standardized quantification with RSEM [9]. They finally obtained $18,354$ samples with $60,498$ genes. Later on, Wang et al. [19] expanded this work by imposing more rigid requirements on the input data. They also used STAR alignment and RSEM quantification but applied a quality control stage between the two and added a batch correction method to the quantification output. This last processing step was meant to eliminate the non-biological effects of data sources. They also eliminated the categories that did not have a counterpart in the other database and ended up having $10,366$ valid samples with approximately $19,000$ genes depending on the tissue.

2.2   Classification/Detection Methods

A handful of works have proposed classification algorithms for cancer using transcriptomic data [2,3,10,12,13,15], however, only the studies by Quinn et al. [14] and Hong et al. [7] have taken into account the necessity to add healthy samples from the GTEx project using the Wang et al. dataset. Quinn et al. fit an anomaly detector to GTEx samples, predict any out-of-distribution TCGA sample as cancerous, and report an accuracy of $>90\%$ in 5/6 of the used tissues. Hong et al. trained two multi-task multilayer perceptrons that classified decease stage, tissue of origin, and neoplastic subclassification in a hierarchical fashion. They used the first $2,000$ principal components of the data as input to their algorithm achieving 99% accuracy in decease state classification, 97% accuracy in tissue of origin classification, and 92% accuracy in neoplastic subclassification.

We compare our classification results with two re-implementations of the Hong et al. model [7]. The original version trained over the first $2,000$ principal components and a second version trained using all genes. To adequate the model to our framework, we implemented just 2 classes (cancer/healthy) in the decease state classification head and lowered the learning rate of the complete feature model by a factor of 100 (for convergence).

We also benchmark our detection results against an adaptation of the original Quinn et al. [14] source code. This detector was trained to detect cancer types instead of healthy tissues in an all-vs-one fashion.

3 Correcting Bias in the Input Data

To formally test for translation biases, we trained a linear support vector classifier (SVC) in both available datasets to predict the data source. The SVCs were trained in 80% of the samples and tested in the remaining 20%. The results can be observed in Table 1. Surprisingly, both data sources (GTEx and TCGA) can be linearly separated in both unified datasets. This observation was expected in the Vivian et al. dataset, as [19] demonstrated the existence of batch effects after standardized quantification, but given that Wang et al. performed batch correction, one would expect the data not to be biased in that case. With these results, the metrics of both the Quinn et al. anomaly detector and the Hong et al. multilayer perceptron lose clinical utility.

Table 1. Weighted average results of a linear support vector classifier predicting the origin of the data (GTEx or TCGA).

Dataset	Normalization	Precision	Recall	F1	Support
Vivian et al. [18]	–	1.00	1.00	1.00	3671
Wang et al. [19]	–	0.99	0.99	0.99	1822
Vivian et al. [18]	Batch	**0.14**	**0.13**	**0.13**	3671

To correct this bias, we performed a simple z-score standardization for each "batch" or origin database. This method centers the data at the origin and imposes a mean of 0 and a standard deviation of 1 for every gene. Considering that the batch correction in Wang et al. did little to un-bias the data and the fact that the Vivian et al. dataset has substantially more samples, we set the former as our working dataframe and perform batch normalization over RSEM $\log_2(TPM + 0.001)$ values. With these changes, a trained SVC (Table 1) could not separate the data sources linearly.

To train and test the models, we filter out $4,892$ genes with no variation over both datasets (standard deviation of 0.0) to remove genes that express exactly the same in all samples. The class distribution of the resulting dataset can be seen in Table 2. Summarizing, we work with $18,354$ samples, $55,602$ genes and make a standard 60/20/20% train/validation/test data partition.

Table 2. Class distribution of the normalized Vivian et al. [18] dataset.

GTEx				TCGA					
Tissue	N	Tissue	N	Class	N	Class	N	Class	N
ADI	517	MUS	396	ACC	77	LIHC	371	UCEC	181
ADR-GLA	131	NER	278	BLCA	407	LUAD	515	UCS	57
BLA	28	OVA	88	BRCA	1098	LUSC	498	UVM	79
BLO	444	PAN	171	CESC	306	MESO	87		
BLO-VSL	606	PIT	107	CHOL	36	OV	427		
BRA	1152	PRO	152	COAD	288	PAAD	179		
BRE	292	SAL-GLA	55	DLBC	47	PCPG	182		
CER	13	SKI	859	ESCA	182	PRAD	496		
COL	359	SMA-INT	92	GBM	165	READ	92		
ESO	666	SPL	100	HNSC	520	SARC	262		
FAL-TUB	5	STO	210	KICH	66	SKCM	469		
HEA	377	TES	165	KIRC	531	STAD	414		
KID	157	THY	338	KIRP	289	TGCT	137		
LIV	169	UTE	91	LAML	243	THCA	512		
LUN	397	VAG	85	LGG	522	THYM	119		

4 Method

4.1 CanDLE

The CanDLE architecture is the simplest approach to the problem using a gradient based method. It is a multinomial logistic regression that given an input $x \in \mathbb{R}^{n_g}$ computes a probability vector $p \in \mathbb{R}^c$ given by:

$$p = \mathrm{softMax}(Wx), \tag{1}$$

where n_g is the number of considered genes, c is the number of classes of the problem and $W \in \mathbb{R}^{c \times n_g}$ is a learnable weight matrix. To perform multilabel classification c is set to the 63 classes available in the Vivian et al. normalized dataset, and to perform all-vs-one detection of an specific class, c is set to 2.

CanDLE is trained with a cross entropy loss. However, given the unbalanced nature of multilabel classification and all-vs-one detection, we weighted the penalization of each class c_i by a δ_i coefficient given by $\delta_i = B/N_i^2$. Where B is a constant set to 2.5×10^5, and N_i is the number of training samples of class c_i.

4.2 Interpretability

An advantage of the simple CanDLE architecture is its interpretation ease. Consider a logit $l_i = w_i x$ associated to the class c_i and computed for a sample x.

Here, w_i is the ith row of W and has one component w_{ij} per gene. Note that the predicted class of x will be the one with the maximum logit. The components of w_i with the highest absolute value are those that influence the most the computation of l_i and, therefore, the classification of x in c_i. To interpret our method, we train CanDLE 100 times with different random partitions and perform a Wald z test [5] for each weight w_{ij} in W ($\alpha = 1 \times 10^{-6}$). We discard non significant weights and use the absolute value of the mean $|\bar{w}_{ij}|$ as a relative importance measure of how much the jth gene influences the prediction of the class c_i.

To obtain a unified list of genes for pan-cancer classification, we select the top $1,000$ genes in each class and order by the number of times that they were selected in any cancer class. We threshold this list at three repetitions (ensuring that chosen genes are important predictors for at least three cancer types) and perform Gene Ontology (GO) biological process enrichment analysis using ShinyGO [6].

Implementation Details: we train our method on an NVIDIA TITAN-X PascalGPU with a learning rate of 10^{-5}, 100 samples per batch and use an Adam [8] optimizer with standard parameters. CanDLE was trained for 20 epochs in multilabel classification and one epoch in all-vs-one detection as it yielded better results. For this same reason, genes with an original standard deviation lower than 0.1 where excluded in the detection task.

5   Results and Discussion

5.1   Mutilabel Classification

Detailed results of cancer and healthy tissue multilabel classification are shown in Table 3. CanDLE is capable of achieving a state of the art performance of 94.1% (test) and 92.0% (validation) balanced accuracy outperforming both versions of the Hong et al. [7] model by a large margin. It obtained an absolute difference of +7.3% (test) and +5.6% (validation) in this metric when compared to the complete feature re-implementation. These results prove that a simple method can correctly discriminate the transcriptomic signatures of all healthy tissue and cancer types even when the biases are removed. Additionally, we observed that, although the Hong et al. model is more complex and flexible, the fact that it performs multiple predictions for a single sample makes the method prone to performing chained errors (i.e. correctly predicting cancerous and colon tissue but erroneously give a kidney cancer subtype classification). Also note that in a clinical setting the multiple predictions offer no benefit over a direct classification performed by CanDLE.

As expected, adding a weighted loss function improved the results in terms of balanced accuracy (Table 3) by +3.2% (test) and +0.8% (validation). This behavior also implied a slight decrease in both total accuracy and mean average precision since the inclusion of weights prioritized a great performance in

Table 3. Multilabel classification results on the normalized Vivian et al. [18] dataset. mACC: balanced accuracy, ACC: accuracy, mAP: mean average precision, PCA: principal component analysis, CF: complete features.

Model	Validation			Test		
	mACC	ACC	mAP	mACC	ACC	mAP
Hong et al. PCA [7]	64.0	65.3	–	61.7	64.9	–
Hong et al. CF [7]	86.4	84.6	–	86.8	84.4	–
CanDLE w/o weights	91.2	**96.0**	**95.3**	90.9	**95.7**	**96.6**
CanDLE	**92.0**	95.6	94.2	**94.1**	95.6	95.0

each class over bulk correct predictions. Such observations suggest that addressing class unbalance using loss weights is an effective technique in the current framework.

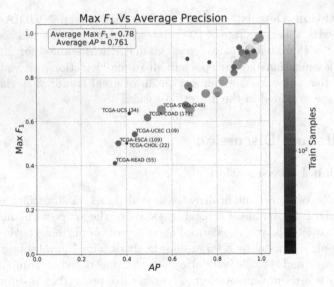

Fig. 1. CanDLE detection max F_1 Vs. AP summary plot for all cancer types over the normalized Vivian et al. [18] validation set. The color and size of each point correspond to the number of training samples available. Specific cancer identifiers are shown for all classes with max F_1 or AP below 0.6. Brackets contain the number of training samples.

5.2 Detection

A summary plot of the all-vs-one CanDLE detection experiments can be seen in Fig. 1. Interestingly, digestive cancers were the hardest to detect. But, not surprisingly, the worst performance was observed in underrepresented cancer types

(READ, CHOL, ESCA, UCEC, UCS, COAD, and STAD). This observation indicates that access to a bigger number of training samples generally helps to obtain better results. However, it is outstanding that CanDLE achieves a mean $\max F_1$ and average AP of 78.0% and 76.1% respectively, considering that in some cases, the number of positive samples was extremely low (e.g., 22 training samples for CHOL). This is especially relevant in clinical practice, where highly unbalanced detection is common.

The adaptation of the Quinn et al. [14] anomaly detector obtained a mean $\max F_1$ of 77.7% and an average AP of 77.2%, making the performance of CanDLE state of the art in mean $\max F_1$ and competitive in average AP. These results are particularly important considering that the Quinn et al. model was explicitly designed for detection tasks while CanDLE is a flexible and simple architecture that can also perform multilabel classification. Moreover, CanDLE has the advantage of providing more transparent and direct interpretability of detection models when compared with the Quinn et al. algorithm.

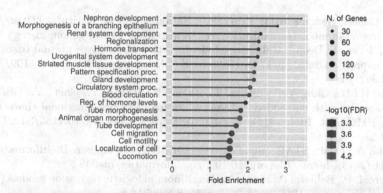

Fig. 2. Biological processes gene ontology results of the top $1,000$ predictor genes shared by at least 3 TCGA classes. Developmental and morphogenesis processes appear to have a mayor role.

5.3 Interpretability

After performing our interpretability protocol, we obtained an ordered list of $1,982$ genes. Notably, the majority of the weights (71.3%) resulted significant in the Wald z test highlighting the reliability of CanDLE. The genes YWHAQ, AC073578, AC013417, and RNU6-1207P were found to be important predictors in more than 10 cancer types. Figure 2 shows the results of GO biological process enrichment analysis where it is clear that CanDLE recognizes developmental and morphogenesis pathways as essential to perform pan-cancer classification.

6 Conclusion

In this work, we empirically prove that previous approaches to joining the TCGA and GTEx databases have significant biases and correct them with a z-score batch standardization. Additionally, we present CanDLE, a simple multinomial logistic regression method that can perform both cancer/healthy tissue type multilabel classification and all-vs-one detection with state of the art performance. Finally, we leveraged the simplicity of CanDLE to interpret gene relevance in pan-cancer classification which recognized developmental and morphogenesis pathways as important predictors.

Acknowledgement. GM acknowledges the support of a UniAndes-DeepMind Scholarship 2022. We also acknowledge the valuable help of Camilo Becerra in graphics and tables preparation, and Danniel Moreno for useful discussions and feedback.

References

1. The Cancer Genome Atlas Program - National Cancer Institute. https://www.cancer.gov/about-nci/organization/ccg/research/structural-genomics/tcga
2. Ahn, T., et al.: Deep learning-based identification of cancer or normal tissue using gene expression data, pp. 1748–1752. IEEE (2018). https://doi.org/10.1109/BIBM.2018.8621108
3. Chen, H.I.H., Chiu, Y.C., Zhang, T., Zhang, S., Huang, Y., Chen, Y.: GSAE: an autoencoder with embedded gene-set nodes for genomics functional characterization. BMC Syst. Biol. **12**(8), 45–57 (2018). https://doi.org/10.1186/S12918-018-0642-2
4. Dobin, A., et al.: STAR: ultrafast universal RNA-seq aligner. Bioinformatics **29**, 15–21 (2013). https://doi.org/10.1093/bioinformatics/bts635
5. Fávero, L.P., Belfiore, P.: Binary and multinomial logistic regression models (2019). https://doi.org/10.1016/B978-0-12-811216-8.00014-8
6. Ge, S.X., Jung, D., Yao, R.: ShinyGO: a graphical gene-set enrichment tool for animals and plants. Bioinformatics **36**, 2628–2629 (2020). https://doi.org/10.1093/bioinformatics/btz931
7. Hong, J., Hachem, L.D., Fehlings, M.G.: A deep learning model to classify neoplastic state and tissue origin from transcriptomic data. Sci. Rep. **12**, 9669 (2022). https://doi.org/10.1038/s41598-022-13665-5
8. Kingma, D.P., Ba, J.L.: Adam: a method for stochastic optimization. In: International Conference on Learning Representations, ICLR 2015 - Conference Track Proceedings (2014). https://arxiv.org/abs/1412.6980v9
9. Li, B., Dewey, C.N.: RSEM: accurate transcript quantification from RNA-Seq data with or without a reference genome. BMC Bioinform. **12**, 323 (2011). https://doi.org/10.1186/1471-2105-12-323
10. Li, Y., et al.: A comprehensive genomic pan-cancer classification using The Cancer Genome Atlas gene expression data. BMC Genomics **18**, 1–13 (2017). https://doi.org/10.1186/S12864-017-3906-0
11. Lonsdale, J., et al.: The genotype-tissue expression (GTEx) project. Nat. Genet. **45**(6), 580–585 (2013). https://doi.org/10.1038/ng.2653
12. Lyu, B., Haque, A.: Deep learning based tumor type classification using gene expression data. bioRxiv p. 364323 (2018). https://doi.org/10.1101/364323

13. Mostavi, M., Chiu, Y.C., Huang, Y., Chen, Y.: Convolutional neural network models for cancer type prediction based on gene expression. BMC Med. Genom. **13**(5), 44 (2020). https://doi.org/10.1186/s12920-020-0677-2

14. Quinn, T.P., Nguyen, T., Lee, S.C., Venkatesh, S.: Cancer as a tissue anomaly: classifying tumor transcriptomes based only on healthy data. Front. Genet. **10**, 599 (2019). https://doi.org/10.3389/fgene.2019.00599

15. Ramirez, R., et al.: Classification of cancer types using graph convolutional neural networks. Front. Phys. **8**, 1–14 (2020). https://doi.org/10.3389/fphy.2020.00203

16. Singh, D., Singh, B.: Investigating the impact of data normalization on classification performance. Appl. Soft Comput. **97**, 105524 (2020). https://doi.org/10.1016/j.asoc.2019.105524

17. Tripathi, R., Sharma, P., Chakraborty, P., Varadwaj, P.K.: Next-generation sequencing revolution through big data analytics. Front. Life Sci. **9**, 119–149 (2016). https://doi.org/10.1080/21553769.2016.1178180

18. Vivian, J., et al.: Toil enables reproducible, open source, big biomedical data analyses. Nat. Biotechnol. **35**, 314–316 (2017). https://doi.org/10.1038/nbt.3772

19. Wang, Q., et al.: Unifying cancer and normal RNA sequencing data from different sources. Sci. Data **5**, 180061 (2018). https://doi.org/10.1038/sdata.2018.61

Cross-Stream Interactions: Segmentation of Lung Adenocarcinoma Growth Patterns

Xiaoxi Pan[1,2], Hanyun Zhang[1,2], Anca-Ioana Grapa[1,2], Khalid AbdulJabbar[1,2], Shan E Ahmed Raza[3], Ho Kwan Alvin Cheung[4], Takahiro Karasaki[4,5], John Le Quesne[6], David A. Moore[5,7], Charles Swanton[4,5,8], and Yinyin Yuan[1,2(✉)]

[1] Centre for Evolution and Cancer, The Institute of Cancer Research, London, UK
yinyin.yuan@icr.ac.uk
[2] Division of Molecular Pathology, The Institute of Cancer Research, London, UK
[3] Department of Computer Science, University of Warwick, Coventry, UK
[4] Cancer Evolution and Genome Instability Laboratory, The Francis Crick Institute, London, UK
[5] Cancer Research UK Lung Cancer Centre of Excellence, UCL Cancer Institute, London, UK
[6] Institute of Cancer Sciences, University of Glasgow, Glasgow, UK
[7] Department of Cellular Pathology, University College London, University College Hospital, London, UK
[8] Department of Medical Oncology, University College London Hospitals NHS Foundation Trust, London, UK

Abstract. Lung adenocarcinoma has histologically distinct growth patterns that have been associated with patient prognosis. Precision segmentation of growth patterns in routine histology samples is challenging due to the complexity of patterns and high intra-class variability. In this paper, we present a novel model with a multi-stream architecture, Cross-Stream Interactions (CroSIn), which fully considers crucial interactions across scales to gather abundant information. The first-order attention introduces contextual information at an early stage to guide low-level feature encoding. The second-order attention then focuses on learning high-level feature relations among scales to extract discriminative features. Experimental results show interactions at both low- and high-level feature learning stages are crucial in performance improvement. The proposed method outperforms state-of-the-art networks, achieving an average Dice of 60.34% at patch level, and an average accuracy of 65.31% at sample level, which is also verified in an independent cohort.

Keywords: Semantic segmentation · Growth patterns · Histology

1 Introduction

Lung adenocarcinoma (LUAD) growth patterns depict the spatial organization of cancer and stromal cells. The wide spectrum of growth patterns, including

W. Qin et al. (Eds.): CMMCA 2022, LNCS 13574, pp. 78–90, 2022.
https://doi.org/10.1007/978-3-031-17266-3_8

micropapillary, solid, papillary, acinar, lepidic, and cribriform, have been recognised by the WHO [25] and several other studies [15,28]. The predominance and composition of growth patterns have been shown to associate with patient survival [19], which highlights the significance of identifying diverse subtypes in the clinical setting. However, it is challenging to distinguish these subtypes without domain language, as growth patterns usually consist of a variety number of cancer cells with different arrangements and morphological structures, and there is no specific shape or size that can be used to quantitatively define them.

Deep learning models have presented notable advantages in histology image analysis, such as tumor segmentation [21,27], nuclei segmentation [8,17], and cell subtype classification [7,12]. Unlike some of these generic applications, tumor growth patterns segmentation present novel challenges to the field, due to their variability and complexity. An accurate delineation of growth pattern-specific masks at a whole-slide image (WSI) level can be achieved with a deep learning approach, such a tool can address the following: 1) empower large-scale, automated and objective analysis to assist current pathological assessment and mitigate inter/intra observer variability; 2) enable a closer look at growth pattern heterogeneity, which can provide an additional insight for patient stratification and prognosis; 3) understand cancer progression with molecular and morphological lenses whilst these patterns evolve from one type to another.

At present, several methods have been proposed to address the growth pattern segmentation problem, which can be essentially grouped into two categories: patch-wise classification [2,10,29] and pixel-wise classification [23,24], also known as semantic segmentation. However, results obtained with existing approaches do not appear to precisely delineate contours, such as alveolar walls in the lepidic pattern and glands in the acinar pattern. Also, all the above methods can only segment 5 growth patterns, leaving out cribriform. Considering its relevance to prognosis and recent classification as high-grade by the International Association for the Study of Lung Cancer (IASLC) [19], we decided to include cribriform as a distinct pattern, and thus develop a segmentation model capable of identifying all 6 recognised patterns.

In this paper, we propose a novel method, Cross-Stream Interactions (CroSIn), to segment 6 types of growth patterns in LUAD. The model fully explores the relations among multiple scales during low- and high-level feature encoding, thereby enriching the feature pool. The motivation behind the proposed method is: 1) a growth pattern is a group behavior of cancer cells, thus contextual information provided by multi-scale could be beneficial for the segmentation; 2) interactions between streams/layers have been proven effective in fine-grained recognition [16,31], therefore we investigate interactions across scales via first- and second-order attentions to achieve a fine-grained segmentation of tumor patterns. The main contribution of our study can be summarized as 3 folds,

- We develop a growth pattern segmentation method by considering cross-scale interactions, which can not only guide low-level feature encoding (first-order attention) by leveraging global information, but also learn high-level feature relations across scales (second-order attention).
- The proposed cross-scale second-order attention module captures common features among scales and their complementary features, corresponding to areas of high and low gradient variations, which reflects morphological characteristics of growth patterns.
- We expand the validation of the proposed method from patch level to WSI level, by correlating predicted subtype percentages with pathological estimations, further demonstrating the effectiveness of the proposed method.

2   Method

The proposed method CroSIn is shown in Fig. 1(a), which encodes three streams, coarse (top path), intermediate (middle) and fine streams (bottom), with different scales of information, and segmentation results are derived from the fine stream. The model applies ResNet50 [13] as the backbone with a number of filters reduced by 4 times (Fig. 1(a)), the down-sampling in the last block of each stream is removed, and the convolution operation is replaced with atrous convolution in order to retain larger receptive fields without adding more parameters. The cross-stream first-order attention (referred to as first-order attention hereafter) shown in Fig. 1(b) is applied to guide low-level feature encoding, while the cross-stream second-order attention (referred to as second-order attention hereafter) Fig. 1(c) aims to direct the relation learning among high-level features obtained from the Pyramid Pooling Module [32]. Such two attention modules are set at different learning stages to thoroughly utilize interactions among streams.

2.1   First-order Attention

Each stream learns different features since the input scale is varied, and it has been shown that features derived from a smaller scale tend to focus on the periphery [4]. In the proposed method, we assume the upper stream with a relatively smaller scale input achieved by an average pooling strategy is able to learn more global features than the other one as it is with a larger receptive field. The receptive field size of the coarse stream is 156×156, which is passed to the intermediate stream at its receptive field size of 102×102. The same with the interaction between intermediate (214×214) and fine streams (147×147). The upper streams learns more global features than the other one, which can guide low-level feature encoding in lower streams. Thus, we apply a top-down attention mechanism between streams to pass global information to the next stream. The coarse stream acts on the intermediate one, which then guides the fine stream learning.

Figure 1(b) shows the details of the first-order attention, which follows the channel and spatial attention proposed in [30]. The difference is that the channel and spatial attention here act on the other scales instead of the same one

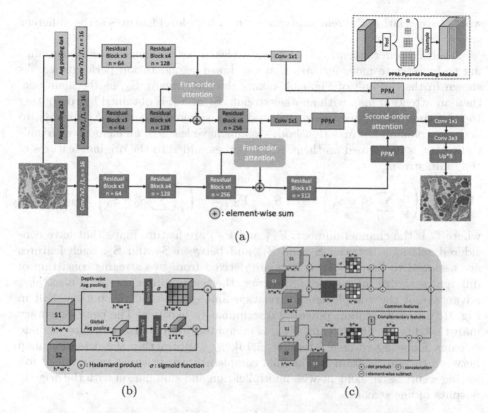

Fig. 1. The architecture of the proposed method CroSIn. (a) Main framework. (b) First-order attention, taking coarse and intermediate streams as instance. (c) Second-order attention.

in the context of multi-scale. The coarse stream, S_1, is firstly performed by depth-wise and global average pooling in parallel followed by a convolution layer and two dense layers, respectively. Then the corresponding attention map can be obtained via a sigmoid function. Lastly, the Hadamard product between the to-be-guided scale, S_2, and channel attention map is computed, whose output then multiplies the spatial attention map in element-wise to deliver the representations. The fine stream learning, S_3, can also be guided similarly, involving contextual information from the intermediate stream, S_2.

2.2 Second-order Attention

As the learning proceeds, high-level features with semantic information are forming. Multiple scales of high-level features have proven to be effective in semantic segmentation for medical images [3,11], while the relation among different scales of high-level features is yet to cover. Furthermore, self-relation (or self-attention) has shown promising performance in segmentation [9,26,33], thus, it is worth investigating the relations concealed in high-level features across different scales.

A second-order attention is proposed to learn the relations across scales, aiming to investigate more features that can boost segmentation performance. As shown in the top half of Fig. 1(c), taking the fine stream, $\mathbf{S}_3$, as the main one, then an attention map with another stream, $\mathbf{S}_1$ or $\mathbf{S}_2$, is obtained by computing the dot product between them followed by a sigmoid function. Afterwards, to activate common features that both streams have learned, an element-wise multiplication is performed on them, which is then added to the original features of the main stream,

$$\mathbf{F}_{3,1} = \sigma\left(\frac{1}{C}\sum_{c=1}^{C} S_1^c \circ S_3^c\right) \circ \mathbf{S}_3 + \mathbf{S}_3, \quad \mathbf{F}_{3,2} = \sigma\left(\frac{1}{C}\sum_{c=1}^{C} S_2^c \circ S_3^c\right) \circ \mathbf{S}_3 + \mathbf{S}_3 \quad (1)$$

where C is the channel number, $\mathbf{F}_{3,1}$ and $\mathbf{F}_{3,2}$ are feature maps that have considered relations between $\mathbf{S}_3$ and $\mathbf{S}_1$, and between $\mathbf{S}_3$ and $\mathbf{S}_2$. Such features are assumed to be reliable as they are derived from two streams consisting of different scales of information. Moreover, the second-order attention also takes advantage of the complementary relation among streams, the bottom half in Fig. 1(c), to fully exploit potential discriminative features. The complementary map that fits for fine stream, $\mathbf{F}_{3,r}$, is computed from coarse and intermediate streams following reverse attention [6,14], $i.e.$, subtracting the attention map between $\mathbf{S}_1$ and $\mathbf{S}_2$ from 1. The final complementary features are obtained following a successive element-wise multiplication and summation with the original outputs of fine stream.

$$\mathbf{F}_{3,r} = \left[1 - \sigma\left(\frac{1}{C}\sum_{c=1}^{C} S_1^c \circ S_2^c\right)\right] \circ \mathbf{S}_3 + \mathbf{S}_3 \quad (2)$$

3 Experimental Results

Experiments were conducted on haematoxylin and eosin (H&E) WSIs of 49 tissue sections collected from the TRACERx 100 cohort [1]. Each WSI was sparsely annotated by the consensus of three senior pathologists. The annotated areas were down-sampled to 10× and then divided into patches with a size of 384 × 384. Accordingly, the experimental dataset consists of 2968 patches covering 6 patterns. An independent cohort consisting of 192 WSIs for 50 cases from LATTICe-A cohort [18] was also used to verify the performance on WSI level.

Pixel-wise Dice coefficient (Macro-F1), F1 score and overall precision (OP), and object-wise Dice (OD) [22] were used as metrics to evaluate the patch-level performance, and the background was excluded. We applied 5-fold cross validation to verify the effectiveness of the proposed method. The average performance across 5 folds was then taken as the estimation of the final performance. Each fold was ensured to consist of all the 6 patterns and remained roughly the same subtype distribution with the whole dataset. WSI-level performance was assessed by Spearman correlation coefficient between predicted proportions and pathological estimations, as well as predominant pattern accuracy.

3.1   Ablation Study

The first- and second-order attention aim to leverage different scales of information at different learning stages, and enable the interactions among streams to expand the feature pool, which could be effective in segmentation of growth patterns. In this section, we systematically verify the proposed model from 2 aspects, 1) effectiveness of multi-stream, and 2) effectiveness of attention modules.

Table 1 left part reports each subtype's segmentation performance measured by Dice coefficient for each method, including baseline method (Single Stream), multi-stream with element-wise add combination (Multi-ADD), multi-stream with only first-order attention (Multi-FO), multi-stream with only second-order attention (Multi-SO) and the proposed CroSIn model (Multi-FO & SO). Overall, the segmentation performance of cribriform and micropapillary is inferior to other patterns in all methods. It is probably due to 1) limited number of training data, and 2) their complex morphological characteristics, e.g. cribriform is an intermediate state of solid and acinar and could appear similar to either of them. In addition, methods with attention modules can achieve the best performance for each subtype, implying that the attention techniques come into effect.

Together with the overall performance shown in the right part of Table 1, multi-stream variants is much more promising than single stream, despite the single stream potentially having an advantage over some of multi-stream methods. Thus, multi-stream can improve the segmentation performance through gathering different types of features. Moreover, utilizing first-order attention (Multi-FO) to combine streams at low-level feature encoding stage can yield better overall performance than simply adding them (Multi-ADD), the F1 score is improved by over 10%. It especially performs well for papillary and lepidic segmentation, both of whose performance are increased by about 10% in terms of Dice, boosting the overall performance. Furthermore, the second-order attention outperforms the first-order attention for most subtype segmentation (except papillary), thereby improving the overall performance by 4.64%, 2.33% and 1.04% regarding Dice, F1 and OP. This suggests that high-level features' interactions across streams could be more effective than merging at low-level feature learning, which is reasonable considering the importance of high-level features in semantic segmentation. The proposed model adopts both first- and second-order attention, further enhancing the overall performance with notable improvements, in particular regarding F1 which is increased by 11%. Therefore, integrating multiple streams via attention modules to enforce an interaction between them can deliver promising performance in growth pattern segmentation. Visual comparison can be found from the Appendix.

The second-order attention maps are visualized in Fig. 2. It can be seen that the attention interacting between streams, Figs. 2 (c) and (d), captures their common sharing features, reflecting the area with high gradient variations, occurring for lepidic (blue) and papillary (yellow). The reason is that these areas are easily distinguishable, so they can be easily captured by different streams, and the multiplication operation between streams in Eq. 1 can

Table 1. Subtype (left, Dice) and overall performance for ablation study(%).

Method	Cri	Mic	Sol	Pap	Aci	Lep	Dice	F1	OP	OD
Single Stream	29.21	44.49	66.37	47.55	56.86	55.39	49.98	34.43	55.59	59.40
Multi-ADD	30.01	43.18	67.93	50.62	61.04	55.43	51.37	38.28	57.04	62.58
Multi-FO	31.50	41.17	66.47	**61.01**	60.74	64.39	54.21	50.53	60.93	66.32
Multi-SO	47.32	**47.33**	**72.78**	54.19	62.53	68.96	58.85	52.86	61.97	65.56
Multi-FO & SO	**49.66**	44.30	71.70	58.63	**63.10**	**74.63**	**60.34**	**63.86**	**65.43**	**69.40**

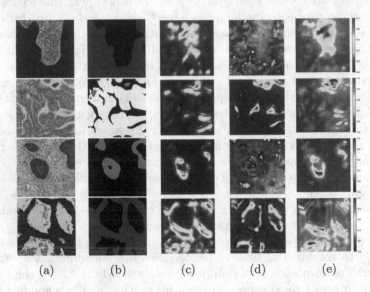

(a)   (b)   (c)   (d)   (e)

Fig. 2. Second-order attention heatmaps, from top to bottom is solid (dark red), papillary (yellow), acinar (red) and lepidic (blue). (a) Ground truth. (b) Results from CroSIn. (c) Attention heatmap interacting between coarse and fine streams. (d) Attention heatmap interacting between intermediate and fine streams. (e) Reverse attention heatmap interacting between coarse and intermediate streams. (Color figure online)

subsequently highlights these areas. The difference between Figs. 2 (c) and (d) also demonstrates the need to consider multi-scale information for growth pattern dissection. The reverse attention in Fig. 2 (e) serves as the complementary focusing on the remaining features, which could be the smooth area due to the subtraction in Eq. 2, such as solid pattern shown in the first row.

3.2   Comparison with State-of-the-Art Methods

In this section, we compare the proposed CroSIn model with representative methods using similar strategies, including attention U-Net [20], DeepLabV3+ [5] (multi-scale features), DANet [9] (self-attention for high-level features) and Medical Transformer [26] (fully self-attention). Table 2 shows the subtype and overall performance of different methods at patch level. It can be seen growth pattern segmentation is a challenging task, which is not only due to their complex heterogeneity but also the variability of annotations collected from different pathologists. Nevertheless, the proposed model achieves the best overall performance regarding the 4 metrics among comparison methods, and DeepLabV3+ is in the second place, indicating multi-stream strategy together with their interactions are necessary in growth pattern segmentation. The attention U-Net failed in segmenting micropapillary, which is probably due to its limited data.

Table 2. Subtype (left, Dice) and overall performance at patch level (%).

Method	Cri	Mic	Sol	Pap	Aci	Lep	Dice	F1	OP	OD
Attention U-Net	18.99	0	71.26	42.69	45.18	48.06	37.70	20.02	50.40	55.36
DeepLabV3+	41.94	41.33	**73.81**	51.86	60.95	68.64	56.91	62.50	62.44	68.45
DANet	23.44	26.13	42.19	34.12	49.71	52.35	37.99	10.90	47.31	44.84
MedT	23.74	12.92	65.92	48.14	41.71	47.86	40.05	36.47	50.96	54.80
CroSIn	**49.66**	**44.30**	71.70	**58.63**	**63.10**	**74.63**	**60.34**	**63.86**	**65.43**	**69.40**

In clinical practice, the predominant pattern and subtype percentages are taken as the score in routine diagnostic slides. Thus, we also compare above methods at WSI level, including Spearman correlation between predicted subtype percentage with its pathological estimation, and predominant pattern accuracy. Given a segmented WSI-level mask, the predominant pattern is the pattern with the maximum number of pixels, and each subtype percentage is computed by dividing the pixel number of a pattern with the total number of pixels covering all the patterns. Higher correlations indicate better performance.

The comparison results on development dataset are shown in Table 3, the proposed method is particularly capable of segmenting solid, papillary and acinar patterns as these patterns' percentages are strongly correlated with pathological estimations ($p < 0.01$). Lepidic and cribriform patterns are moderately correlated with the ground truth, but still outperforming all other methods. In regards to micopapillary, the performance is limited and weakly correlated. Overall, this comparison on predominant pattern accuracy and parameters further demonstrate the advantages of the proposed method over previous work. It should be noted that the evaluation at WSI level is also conducted across 5 folds.

The comparison on an independent cohort, Table 4, also shows advantages of the proposed method over other algorithms regarding accuracy and subtype prediction. The correlation of micropapillary drops drastically for all the methods, which is probably due to the data bias. Visualisation of patch-level and WSI-level masks can be found from the Appendix.

Table 3. Performance comparison at WSI level for the development dataset. NaN indicates the lack of model prediction for the pattern.

Method	Cri	Mic	Sol	Pap	Aci	Lep	Accuracy (%)	Paras
Attention U-Net	0.3108	NaN	**0.7798**	0.3085	0.5163	0.2828	48.98	15.55 M
DeepLabV3+	0.3320	0.2790	0.7613	0.5321	0.6490	0.4817	55.10	11.85 M
DANet	0.0994	0.2290	0.5278	0.0282	0.4959	0.1191	42.86	6.67 M
MedT	0.1659	−0.1232	0.6346	0.3685	0.5853	0.2792	42.86	1.41 M
CroSIn	**0.4244**	**0.3026**	0.7545	**0.6237**	**0.6885**	**0.5639**	**65.31**	4.10 M

Table 4. Performance comparison at WSI level for the independent cohort.

Method	Cri	Mic	Sol	Pap	Aci	Lep	Accuracy (%)
Attention U-Net	0.4443	NaN	0.6175	−0.0012	0.1970	0.1443	42.00
DeepLabV3+	0.2795	0.1134	0.1430	0.0916	0.1459	0.1137	32.00
DANet	0.2182	−0.0496	0.4071	0.1735	0.5169	**0.4187**	26.00
MedT	0.3700	**−0.2606**	0.1997	0.3931	0.1284	0.1186	16.00
CroSIn	**0.5160**	0.0535	**0.7840**	**0.5319**	**0.6317**	0.2450	**60.00**

4　Conclusion

In this paper, we have proposed a segmentation method, CroSIn, for growth pattern segmentation in LUAD by leveraging cross-stream interactions. First- and second-order attention modules can fully consider relations among streams at different stages, resulting in an expanded feature pool and allowing the model to capture discriminative features, crucial to decipher tumor growth patterns. Experimental results at patch level verify the effectiveness of the first- and second-order attentions in segmentation, implying that the hidden interactions among streams cannot be ignored. At both patch and WSI levels on either development or independent dataset, our method outperforms state-of-the-art methods and offers several advantages. When incorporated with other phenotypes, CroSIn will allow the spatial tracking of ecological hallmarks to better understand lung cancer evolution.

Appendix

(Figures 3, 4, 5 Table 5).

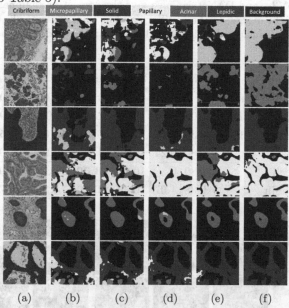

Fig. 3. Instances of segmentation results for ablation study, showing the effectiveness of the first- and second-order attention modules. (a) Ground truth. (b) Single Stream. (c) Multi-ADD. (d) Multi-FO. (e) Multi-SO. (f) Multi-FO & SO.

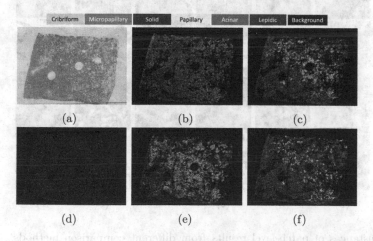

Fig. 4. Segmentation instances at WSI level from different comparison methods. (a) Original WSI with acinar as predominant pattern (red). (b) attention U-Net. (c) DeepLabV3+. (d) DANet. (e) Medical Transformer. (f) Proposed CroSIn. (Color figure online)

Table 5. Performance comparison at WSI level via predominant pattern and subtype percentages (%) for results in Fig. A2, and the bold text indicates predominant pattern. The result obtained from CroSIn is in line with ground truth in terms of predominant pattern. DeepLabV3+ and DANet can also give the correct predominant pattern, acinar, but with a slight margin to the papillary and lepidic, respectively. Both attention U-Net and Medical Transformer (MedT) yield papillary as predominant pattern, which are mispredicted.

Method	Cri	Mic	Sol	Pap	Aci	Lep
Ground truth	0	0	0	0	**90.00**	10.00
Attention U-Net	1.09	0	2.42	**45.10**	33.09	18.29
DeepLabV3+	0.33	5.99	1.77	36.76	**53.81**	1.35
DANet	0.80	0.24	10.44	11.93	**42.29**	34.30
MedT	1.80	0	10.08	**63.00**	22.32	2.80
CroSIn	0.38	1.07	0.48	21.69	**74.32**	2.07

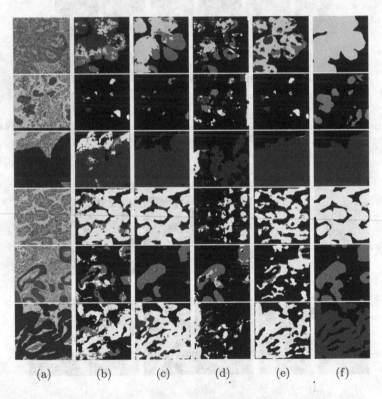

(a)   (b)   (c)   (d)   (e)   (f)

Fig. 5. Instances of patch-level results from different comparison methods, suggesting the advantage of the proposed model. (a) Ground truth. (b) attention U-Net. (c) DeepLabV3+. (d) DANet. (e) Medical Transformer. (f) Proposed CroSIn.

References

1. AbdulJabbar, K., Raza, S.E.A., Rosenthal, R., Jamal-Hanjani, M., Veeriah, S., et al.: Geospatial immune variability illuminates differential evolution of lung adenocarcinoma. Nat. Med. **26**(7), 1054–1062 (2020)
2. Alsubaie, N., Shaban, M., Snead, D., Khurram, A., Rajpoot, N.: A multi-resolution deep learning framework for lung adenocarcinoma growth pattern classification. In: Nixon, M., Mahmoodi, S., Zwiggelaar, R. (eds.) MIUA 2018. CCIS, vol. 894, pp. 3–11. Springer, Cham (2018). https://doi.org/10.1007/978-3-319-95921-4_1
3. Chen, H., Qi, X., Yu, L., Heng, P.A.: Dcan: deep contour-aware networks for accurate gland segmentation. In: CVPR, pp. 2487–2496 (2016)
4. Chen, L.C., Yang, Y., Wang, J., Xu, W., Yuille, A.L.: Attention to scale: scale-aware semantic image segmentation. In: CVPR, pp. 3640–3649 (2016)
5. Chen, L.C., Zhu, Y., Papandreou, G., Schroff, F., Adam, H.: Encoder-decoder with atrous separable convolution for semantic image segmentation. In: ECCV, pp. 801–818 (2018)
6. Chen, S., Tan, X., Wang, B., Hu, X.: Reverse attention for salient object detection. In: ECCV, pp. 234–250 (2018)
7. Coudray, N., Ocampo, P.S., Sakellaropoulos, T., Narula, N., Snuderl, M., et al.: Classification and mutation prediction from non-small cell lung cancer histopathology images using deep learning. Nat. Med. **24**(10), 1559–1567 (2018)
8. Feng, Z., Wang, Z., Wang, X., Mao, Y., Li, T., et al.: Mutual-complementing framework for nuclei detection and segmentation in pathology image. In: CVPR, pp. 4036–4045 (2021)
9. Fu, J., Liu, J., Tian, H., Li, Y., Bao, Y., et al.: Dual attention network for scene segmentation. In: CVPR, pp. 3146–3154 (2019)
10. Gertych, A., Swiderska-Chadaj, Z., Ma, Z., Ing, N., Markiewicz, T., et al.: Convolutional neural networks can accurately distinguish four histologic growth patterns of lung adenocarcinoma in digital slides. Sci. Rep. **9**(1), 1–12 (2019)
11. Graham, S., Chen, H., Gamper, J., Dou, Q., Heng, P.A., et al.: Mild-net: minimal information loss dilated network for gland instance segmentation in colon histology images. Med. Image Anal. **52**, 199–211 (2019)
12. Hashimoto, N., Fukushima, D., Koga, R., Takagi, Y., Ko, K., et al.: Multi-scale domain-adversarial multiple-instance CNN for cancer subtype classification with unannotated histopathological images. In: CVPR, pp. 3852–3861 (2020)
13. He, K., Zhang, X., Ren, S., Sun, J.: Deep residual learning for image recognition. In: CVPR, pp. 770–778 (2016)
14. Huang, Q., Xia, C., Wu, C., Li, S., Wang, Y., et al.: Semantic segmentation with reverse attention. In: BMVC (2017)
15. Kadota, K., Yeh, Y.C., Sima, C.S., Rusch, V.W., Moreira, A.L., et al.: The cribriform pattern identifies a subset of acinar predominant tumors with poor prognosis in patients with stage i lung adenocarcinoma: a conceptual proposal to classify cribriform predominant tumors as a distinct histologic subtype. Mod. Pathol. **27**(5), 690–700 (2014)
16. Lin, T.Y., RoyChowdhury, A., Maji, S.: Bilinear CNN models for fine-grained visual recognition. In: ICCV, pp. 1449–1457 (2015)
17. Liu, D., Zhang, D., Song, Y., Zhang, F., O'Donnell, L., et al.: Unsupervised instance segmentation in microscopy images via panoptic domain adaptation and task reweighting. In: CVPR, pp. 4243–4252 (2020)

18. Moore, D.A., et al.: In situ growth in early lung adenocarcinoma may represent precursor growth or invasive clone outgrowth-a clinically relevant distinction. Mod. Pathol. **32**(8), 1095–1105 (2019)
19. Moreira, A.L., Ocampo, P.S., Xia, Y., Zhong, H., Russell, P.A., et al.: A grading system for invasive pulmonary adenocarcinoma: a proposal from the international association for the study of lung cancer pathology committee. J. Thorac. Oncol. **15**(10), 1599–1610 (2020)
20. Oktay, O., Schlemper, J., Folgoc, L.L., Lee, M., Heinrich, M., et al.: Attention u-net: learning where to look for the pancreas. In: MIDL (2018)
21. Qaiser, T., Tsang, Y.W., Taniyama, D., Sakamoto, N., Nakane, K., et al.: Fast and accurate tumor segmentation of histology images using persistent homology and deep convolutional features. Med. Image Anal. **55**, 1–14 (2019)
22. Sirinukunwattana, K., Pluim, J.P.W., Chen, H., Qi, X., Heng, P., et al.: Gland segmentation in colon histology images: the glas challenge contest. Med. Image Anal. **35**, 489–502 (2017)
23. Tokunaga, H., Iwana, B.K., Teramoto, Y., Yoshizawa, A., Bise, R.: Negative pseudo labeling using class proportion for semantic segmentation in pathology. In: Vedaldi, A., Bischof, H., Brox, T., Frahm, J.-M. (eds.) ECCV 2020. LNCS, vol. 12360, pp. 430–446. Springer, Cham (2020). https://doi.org/10.1007/978-3-030-58555-6_26
24. Tokunaga, H., Teramoto, Y., Yoshizawa, A., Bise, R.: Adaptive weighting multi-field-of-view CNN for semantic segmentation in pathology. In: CVPR, pp. 12597–12606 (2019)
25. Travis, W.D., Brambilla, E., Nicholson, A.G., Yatabe, Y., Austin, J.H., et al.: The 2015 world health organization classification of lung tumors: impact of genetic, clinical and radiologic advances since the 2004 classification. J. Thorac. Oncol. **10**(9), 1243–1260 (2015)
26. Valanarasu, J.M.J., Oza, P., Hacihaliloglu, I., Patel, V.M.: Medical transformer: gated axial-attention for medical image segmentation. In: MICCAI (2021)
27. Wang, X., Fang, Y., Yang, S., Zhu, D., Wang, M., et al.: A hybrid network for automatic hepatocellular carcinoma segmentation in H&E-stained whole slide images. Med. Image Anal. **68**, 101914 (2021)
28. Warth, A., Muley, T., Kossakowski, C., Stenzinger, A., Schirmacher, P., et al.: Prognostic impact and clinicopathological correlations of the cribriform pattern in pulmonary adenocarcinoma. J. Thorac. Oncol. **10**(4), 638–644 (2015)
29. Wei, J.W., Tafe, L.J., Linnik, Y.A., Vaickus, L.J., Tomita, N., Hassanpour, S.: Pathologist-level classification of histologic patterns on resected lung adenocarcinoma slides with deep neural networks. Sci. Rep. **9**(1), 1–8 (2019)
30. Woo, S., Park, J., Lee, J.Y., Kweon, I.S.: CBAM: convolutional block attention module. In: ECCV, pp. 3–19 (2018)
31. Yu, C., Zhao, X., Zheng, Q., Zhang, P., You, X.: Hierarchical bilinear pooling for fine-grained visual recognition. In: ECCV, pp. 574–589 (2018)
32. Zhao, H., Shi, J., Qi, X., Wang, X., Jia, J.: Pyramid scene parsing network. In: CVPR, pp. 2881–2890 (2017)
33. Zhou, Y., Chen, H., Xu, J., Dou, Q., Heng, P.-A.: IRNet: instance relation network for overlapping cervical cell segmentation. In: Shen, D., et al. (eds.) MICCAI 2019. LNCS, vol. 11764, pp. 640–648. Springer, Cham (2019). https://doi.org/10.1007/978-3-030-32239-7_71

Modality-Collaborative AI Model Ensemble for Lung Cancer Early Diagnosis

Wanxing Xu[1], Yinglan Kuang[2], Lin Wang[3], Xueqing Wang[3], Qiaomei Guo[3], Xiaodan Ye[4], Yu Fu[2], Xiaozheng Yang[2], Jinglu Zhang[2,5,6], Xin Ye[2], Xing Lu[2(✉)], and Jiatao Lou[3(✉)]

[1] School of Medicine, Jiangsu University, Zhenjiang, Jiangsu, China
[2] Zhuhai Sanmed Biotech Ltd. Zhuhai, Guangzhou, China
lv.xing@sanmedbio.com
[3] Department of Laboratory Medicine, Shanghai General Hospital, Shanghai, China
loujiatao@sjtu.edu.cn
[4] Department of Radiology, Zhongshan Hospital, Shanghai, China
[5] State Key Laboratory of Quality Research in Chinese Medicine, University of Macau, Macau, China
[6] Institute of Chinese Medical Sciences, University of Macau, Macau, China

Abstract. It is imperative to predict pulmonary nodule malignancy as CT scans become more popular and cancer early detection has become widely recognized for lung cancer detection in its early stages, which could significantly prolong patient survival. Our study compared multi-modality models for the early detection of lung cancer, including traditional diagnostic models and deep learning based LDCT AI models. Furthermore, a multi-model, multi-modality ensemble classifier based on the random forest is also proposed and tested in this study. AUCs of 0.694 and 0.785 were achieved by two CT Image AI models, respectively, in the test clinical cohort consisting of 177 patient CT scans. Based on an ensemble of Random Forest-based multi-modality models combining CT AI models and clinical data, the AUC performance was further improved to 0.846.

Keywords: Early lung cancer diagnosis · Low-dose CT · Artificial intelligence · Cancer diagnostic model · Model ensemble

1 Introduction

Lung cancer is the second cause of cancer incidence and the leading cause of cancer mortality worldwide. GLOBOCAN data estimated that more than 2.2 million newly diagnosed cases and approximately 1.8 million deaths were caused by lung cancer in 2020, accounting for 11.4% and 18% of the global cancer incidence and mortality [1].

W. Xu and Y. Kuang—Contributes equally to this study.

Supplementary Information The online version contains supplementary material available at https://doi.org/10.1007/978-3-031-17266-3_9.

Early detection is critical to improving the lung cancer survival rate. The 5-year survival rate of stage IA lung cancer is 82%, but only 36% and 6% when in stage IIIA and IV [2]. Low dose computed tomography (LDCT) has been proved to reduce 20% lung cancer mortality compared with Chest X-ray by the National Lung Screening Trial (NLST) and has become the only recommended test for lung cancer screening [3].

However, the false-positive rate of LDCT has diminished its efficacy, as only 3.6% of people with positive LDCT results were confirmed to be cancer in the NLST study. Therefore, clinicians utilize decision-making tools in concurrence with LDCT to stratify patients' malignancy risk. Initially, predictive models that consisted of clinical characteristics such as age, gender, family history of cancer, etc., were widely applied in the clinical setting to evaluate individuals' lung cancer risk. The Mayo Clinic Model developed in 1997 by the Mayo Clinic and the Brock Model generated from the PanCan data set in 2013 are commonly found in the clinical setting assisting physicians to determine the probability of lung cancer in pulmonary nodules [4, 5].

Nowadays, Artificial intelligence (AI) approaches have received attention for image analysis in the clinical setting. The use of AI can help clinicians interpret the variations in pathology results and reduce the risk of potential fatigue caused by classifying large numbers of medical images [6], which can improve the diagnostic efficacy and accuracy of lung cancer screening [7]. Therefore, AI is considered a cancer diagnosis tool with great promise. In addition, deep learning technology is capable to improve the effectiveness of lung cancer diagnosis by precisely distinguishing malignant and benign nodules by interpreting specific features and complex patterns from medical images [8]. Using AI algorism as supplementary tools for early lung cancer diagnosis has been adopted in the clinical setting, some AI models performed equal or even more accurate than skilled clinicians in identifying benign from malign pulmonary nodules, which can enhance the diagnostic accuracy and reduce the false-positive rate of LDCT [9].

In this study, we assessed both an in-house developed AI and a commercially available AI for early lung cancer diagnosis and compared their performances with several conventional clinical models. Additionally, we found that combining those solutions with a multi-modality collaborative AI ensemble might improve the accuracy of distinguishing benign from malignant pulmonary nodules.

2   Method

2.1   Study Design and Participants

A total of 480 participants were included between November 2016 and May 2018, from Shanghai Chest Hospital. Written informed consent was obtained from each participant. The study was approved by the ethics committee (ethical approval number KS1961) and registered in the Chinese Clinical Trial Registry (ChiCTR2000036938). The lung cancer diagnosis was confirmed by histopathology according to the guidelines of the National Comprehensive Cancer Network which consists of 123 benign and 303 malignant. Lung cancer staging was performed based on the 8th IASLC TNM Staging System. Lung benign diseases included pneumonia, chronic obstructive pulmonary disease, tuberculosis, and others.

The patients were enrolled according to certain inclusion and exclusion criteria. The inclusion criteria include: 1) contain lung window scan with slice thickness less than 2 mm; 2) without any NULL values for all necessary information. The exclusion criteria were: 1) without lung window scan; 2) only contain slice thickness more than 2 mm; 3). Image quality problem that does not pass the examination by human eyes; 4). Contain Null values.

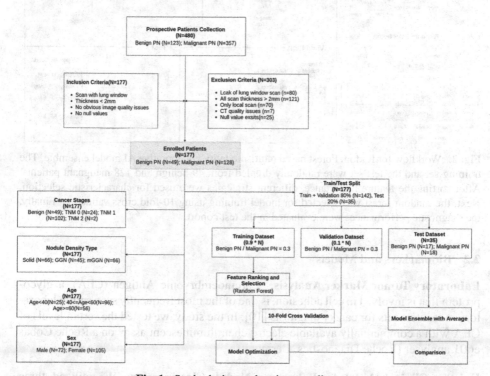

Fig. 1. Study design and patient enrollment.

Figure 1 illustrated the study cohort, which comprise 177 patients, 49 of whom had benign lung cancer and 128 of whom had malignant lung cancer. A number of clinically available biomarkers and models along with two LDCT AI models will be enrolled for overall comparisons and subgroup analyses in which stage, age, sex, and density types will be considered. As shown in Fig. 2, all enrolled patients were undergoing both sub-group analysis and comparison studies, and a Random-forest model was trained to ensemble different information to obtain a better performance, to fully ensure the different modalities and different image AI models.

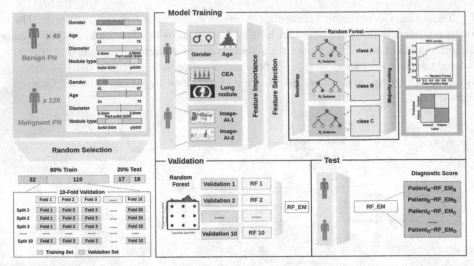

Fig. 2. Workflow for Radom Forest based multi-modality collaborative AI model ensemble. The training set and the test set were randomly divided from 49 benign and 128 malignant patients. After sorting the feature importance, different strategies were used for characteristic selection. Next, the random forest was selected for model training using 10-fold cross validation. Finally, the diagnostic performances were evaluated in the test cohort.

2.2 Biomarkers and Models

Laboratory Tumor Marker Analysis. Carcinoembryonic Antigen (CEA), a glyco-protein that is involved in cell adhesion, is one of the most frequently seen serum labora-tory measurements for cancer diagnosis [10]. In the study, we tested the serum levels of CEA with a commercially available electro chemiluminescent assay on a Roche Cobas e601 analyzer (Roche Diagnostics, Germany).

Existing Clinical Models for Pulmonary Nodule Classification. We utilized three clinical prediction models and measured their diagnosis power in our study cohort: the Mayo clinical model, the Brock Model, and the (Veterans Affairs) VA model. The Mayo clinical model was developed based on a cohort of 629 patients with newly dis-covered 4–30 mm indeterminate solid pulmonary nodules (SPNs). The perdition model has involved characteristics of age, cigarette smoking status, history of extrathoracic cancer, nodule diameter, spiculation, and upper lobe position [4]. The Brock model was established from 1871 persons with 7008 nodules in the PanCan data set and 1090 persons with 5021 nodules in the BCCA data set. Other than clinical characteristics of age, sex, family history of cancer, history of emphysema, radiological characteristics of nodule type, nodule location, and nodule count per scan were also included [5]. The VA model was built from 375 U.S. veterans with SPNS and contained risk factors of smoking history, age, larger nodule diameter, and time since quitting smoking [11].

LDCT Image Artificial Intelligence. CT scan data were obtained with a 128-detector row scanner (Brilliance, Philips, Cleveland, OH, USA) using the helical technique at the end of inspiration during one breath-hold. The scanning parameters of routine CT were as follows: pitch, 1.0; matrix,1024 × 1024; FOV, 300 mm; 120 kVp and 200 mA.

In this study, two deep convolutional neural network model-based artificial intelligence software (Image-AI) were used to detect and classify lung nodules, including Image AI_1 (Deepwise, Beijing, China), and Image AI_2 (Sanmed Biotech, Zhuhai, China). Image-AI-1 has been approved by the NMPA as Level III Medical Device only for pulmonary nodule detection but is equipped with a nodule malignancy prediction module. The Image AI_2 was developed in-house and the performance of detecting nodules and predicting malignancy has also been validated in a large-scale clinical study [12]. To the best of our knowledge, no regulatory body has approved an AI-based nodule malignancy prediction algorithm. All the two AI models have predictions on the malignant degree of the biggest nodule found in the patient, which is also further confirmed by clinicians. The value of this prediction ranges from 0 to 1, with a higher value indicating a higher likelihood of malignancy.

2.3 Multi-modality Multi-model Ensemble

This study employed a Random Forest (RF) based model ensemble strategy to fully leverage all the markers and models. As the tree-based strategy used by random forests naturally that rank by how well the features improve the purity of the node, it is commonly used for feature selection. Following the features selections with the RF, we trained an ensemble RF model using all selected features and the top two features to assess how much the performance improved. The enrolled patients were split into 8/2 as training and testing datasets. The training dataset was 10-fold cross-validated to enhance model generalization, while the test dataset was compared with AUC as a performance metric.

2.4 Model Performance Evaluation

In this initial study, receiver operating characteristic (ROC) curve is used to evaluate the performance of all the biomarkers and models in the prediction of the nodule. The area under Curve (AUC) value is used for comparison between different models and model ensemble strategies.

3   Results

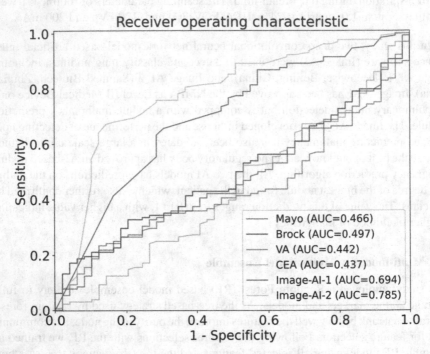

Fig. 3. Receiver operating characteristic of different models.

In Fig. 3, it is evident that both Image-AIs achieved significantly higher AUC scores than the traditional clinical models and biomarkers. The Image-AI-2 is the most accurate of the two Image-AIs, scoring 0.785, slightly better than Image-AI-1's 0.694.

The various study subgroups are shown in Fig. 4. Image-AI-2 has the best performance in most of the subgroups. The performance of the tested models is not significantly affected by gender. Using age as a threshold of 40 and 60, we divided the cohort into three subgroups. For groups below 40 and 40 ~ 60, image-AI-2 has reached AUC scores of 0.855 and 0.749. Image-AI-2 has an AUC score of 0.824 for the subgroup above 60 years old. Image-AI-2 performs slightly better than Image-AI-1 in TNM 2 cancer, while Image-AI-1 outperforms all other models in earlier cancer stages. When subgroup analysis of nodule subtypes, including Solid, GGN (ground glass nodule), and PSN (part-solid nodule), was performed, Image-AI-1 provides the highest AUC score of 0.752 in the Solid groups. On the other hand, Image-AI-2 performs better in GGN, with an AUC of 0.872, which is significantly greater than Image-AI-1's 0.709.

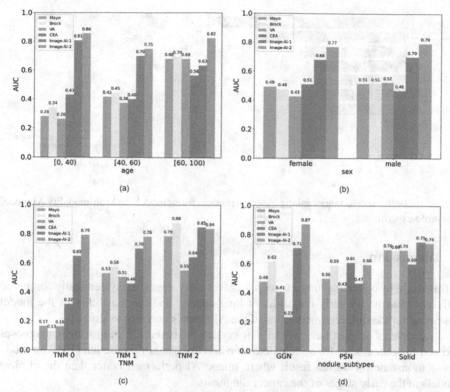

Fig. 4. Subgroup analysis for patient age, sex, lung cancer stage, and lung nodule type. (a) Three age cohorts: under 40, 40–60, and 60–100. (b) subgroups divided by gender. (c) The malignant cohort of different disease stages: stage 0, stage I, and stage II. (d) subgroups with different nodule types: solid nodule, GGN (ground glass nodule), and PSN (part-solid nodule).

Figure 5 shows the results for model ensembles with different strategies. In the predictions model built from the mean value of the two models, the AUC is 0.799, slightly higher than Image-AI-2, the individual predictor with the best performance. Among the different tested machine learning models, the Random Forest model stands out as the final ensemble model. Detailed performance comparison can be found in supplementary Table 1. Using the Random-Forest based model ensemble for the two Image-AI, the AUC is improved to 0.801. The features importance of the Random-forest is described in Fig. 5(a), where Image-AI-2 is the most influential factor, followed by nodule subtypes. The ROC curves in Fig. 5(b) shows that the ensemble model with all features achieves the highest AUC of 0.846, while only utilizing the top two features (Image-AI-2, and nodule subtype) would slightly lower it to 0.833.

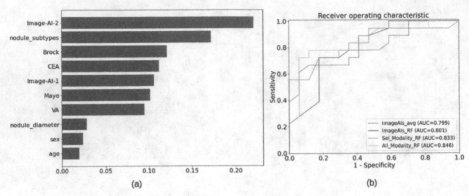

Fig. 5. (a) Feature importance ranking based on random forest model. (b) Multimodality AI model ensemble experiments.

4 Discussion

In this study, by collecting lung cancer patients at early stages, we systemically compared different models, for early diagnosis of lung cancer, and the feasibility of the model ensemble for the different modalities of the AI models was also evaluated.

A couple of studies have stated that certain AI-based CT lung cancer diagnoses showed high accuracies for nodule detection and malignancy prediction [13]. Our study has demonstrated a similar result where image AI performed better than the clinical models at the early stages of the cancer diagnosis.

However, in real-world data tests, all image AI models perform below 0.8 AUC in the early stages. The best performing model is Image-AI-2, provided by Sanmed Biotech Ltd., with an AUC score of 0.79. The image-AI-1 has only achieved an AUC of 0.69. As we carefully control the input image quality, both image AIs can be affected by the image quality. In the sub-group studies, the image-AI-2 and image-AI-1 each has their advantages, which may mainly be due to the dataset bias for the training stage of the AIs. For example, in the sub-group analysis, image-AI-1 has superior performance in the solid type nodules, whereas obtaining deprived performance in the PSN subgroup than the image-AI-2.

The Model ensemble would often improve the final performance of a classifier. By combining Image AIs with clinical models, the model ensemble improved the final score from 0.79 to 0.846. In an ensemble RF model, using multiple image AI models could improve performance by minimizing training dataset bias, and more information, such as clinical results, could improve performance even further. During the training procedure, different models gather datasets with different biases, so the data might be better predicted via the ensemble of models.

In this study, there are a few drawbacks. For a comparative evaluation of the imaging-AI models, we only considered CT scans that were less than 2 mm thick to ensure accurate detection of small nodules. Though the thin-thickness CT scan for lung cancer screening is already the most popular CT scan in most hospitals globally, thick scans are still being performed in some under-equipped hospitals. The other drawback is that all patient data come from one study site, a multi-center study would be investigated in the future.

5 Conclusion

Our study examines the performance of CT AI on real-world lung cancer patient data, as well as other clinical models. While the CT AI algorithms outperformed all other existing diagnostic models, their overall malignancy prediction AUCs were below 0.8, which might not be adequate for clinical applications. Unless massively large input data covering the variety and heterogeneity of lung cancer are available, it is prohibitively costly and difficult to overcome the demography bias of individual models. This study presents a multimodality collaborative AI ensemble that increases the accuracy of lung nodule classifiers, which has greater potential for real-world implementation.

Acknowledgements. This work was supported by grants from Innovation Group Project of Shanghai Municipal Health Commission (2019CXJQ03), the National Natural Science Foundation of China (82002224), Shanghai Municipal Health Commission (20204Y0202).

References

1. Sung, H., Ferlay, J., Siegel, R.L., et al.: Global cancer statistics 2020: GLOBOCAN estimates of incidence and mortality worldwide for 36 cancers in 185 countries. Cancer J. Clin. **71**(3), 209–249 (2021)
2. Goldstraw, P., Chansky, K., Crowley, J., et al.: The IASLC lung cancer staging project: proposals for revision of the TNM stage groupings in the forthcoming (eighth) edition of the TNM classification for lung cancer. J. Thorac. Oncol. **11**(1), 39–51 (2015)
3. The National Lung Screening Trial Research Team. Reduced lung-cancer mortality with low-dose computed tomographic screening. N. Engl. J. Med. **365**(5) 395409 (2011)
4. Swensen, S.J.: The probability of malignancy in solitary pulmonary nodules. Arch. Intern. Med. **157**(8), 849–855 (1997)
5. McWilliams, A., Tammemagi, M.C., Mayo, J.R., et al.: Probability of cancer in pulmonary nodules detected on first screening CT. N. Engl. J. Med. **369**(10), 910–919 (2013)
6. Shen, D., Wu, G., Suk, H.: Deep learning in medical image analysis. Annu. Rev. Biomed. Eng. **19**(1), 221–248 (2017)
7. Goryński, K., et al.: Artificial neural networks approach to early lung cancer detection. Central Eur. J. Med. **9**(5), 632–641 (2014). https://doi.org/10.2478/s11536-013-0327-6
8. Ardila, D., Kiraly, A.P., Bharadwaj, S., et al.: End-to-end lung cancer screening with three-dimensional deep learning on low-dose chest computed tomography. Nat. Med. **25**, 954–961 (2019)
9. Espinoza, J.L.: Artificial intelligence tools for refining lung cancer screening. J. Clin. Med. **9**(12), 3860 (2020)
10. Robert, H.F.: Carcinoembryonic antigen. Ann. Intern. Med. **104**(1), 66–73 (1986)
11. Michael, K.G., Lakshmi, A., Paul, G.B.: A clinical model to estimate the pretest probability of lung cancer in patients with solitary pulmonary nodules. Chest **131**(2), 383–388 (2007)
12. Xinguan, Y., Jianxing, H., Wang, J., et al.: CT-based radiomics signature for differentiating solitary granulomatous nodules from solid lung adenocarcinoma. Lung Cancer **125**, 109–114 (2018)
13. Roger, Y.K., Jason, L.O., Lyndsey, C.P., et al.: Artificial intelligence tool for assessment of indeterminate pulmonary nodules detected with CT. Radiology **000**, 1–9 (2022)

Clustering-Based Multi-instance Learning Network for Whole Slide Image Classification

Wei Wu[1], Zhonghang Zhu[1], Baptiste Magnier[2], and Liansheng Wang[1(✉)]

[1] Department of Computer Science at School of Informatics, Xiamen University, Xiamen, China
{weiwu,zzhonghang}@stu.xmu.edu.cn, lswang@xmu.edu.cn
[2] Euromov Digital Health in Motion, Univ Montpellier, IMT Mines Ales, Ales, France
baptiste.magnier@mines-ales.fr

Abstract. Automated and accurate classification of Whole Slide Image (WSI) is of great significance for the early diagnosis and treatment of cancer, which can be realized by Multi-Instance Learning (MIL). However, the current MIL method easily suffers from over-fitting due to the weak supervision of slide-level labels. In addition, it is difficult to distinguish discriminative instances in a WSI bag in the absence of pixel-level annotations. To address these problems, we propose a novel Clustering-Based Multi-Instance Learning method (CBMIL) for WSI classification. The CBMIL constructs feature set from phenotypic clusters to augment data for training the aggregation network. Meanwhile, a contrastive learning task is incorporated into the CBMIL for multi-task learning, which helps to regularize the feature aggregation process. In addition, the centroid of each phenotypic cluster is updated by the model, and the weights of the WSI patches are calculated by their similarity to the phenotypic centroids to highlight the significant patches. Our method is evaluated on two public WSI datasets (CAMELYON16 and TCGA-Lung) for binary tumor and cancer sub-types classification and achieves better performance and great interpretability compared with the state-of-the-art methods. The code is available at: https://github.com/wwu98934/CBMIL.

Keywords: Whole slide image · Multiple instance learning · Multi-task

1 Introduction

Whole Slide Images (WSIs) which are digital visualization of tissue section are widely used in disease diagnosis [5,22]. Recently, deep learning approaches have been used in WSI analysis, which is a long-standing challenge due to the gigapixel resolution and the lack of pixel-level annotations [24]. Therefore, the analysis of WSI which is a weakly supervised learning problem usually follows a MIL problem formulation [7,20], where each WSI is regarded as a bag containing many instances that are patches of the WSI.

© The Author(s), under exclusive license to Springer Nature Switzerland AG 2022
W. Qin et al. (Eds.): CMMCA 2022, LNCS 13574, pp. 100–109, 2022.
https://doi.org/10.1007/978-3-031-17266-3_10

In previous MIL approaches for WSI analysis, a WSI has been tiled into a large number of small patches and further extracted into features by a pre-trained Convolutional Neural Network (CNN) *e.g.*, ResNet-18 [11]. Then, patch-level features are aggregated, and examined by a classifier that predicts slide-level labels. For aggregation operator, a straightforward method is named pooling, such as mean-pooling and max-pooling [8,13,27]. However, the pooling operation is a handcrafted method that guides limited performance. To address this problem, Ilse *et al.* [12] proposed an attention-based aggregation operator parameterized by deep neural networks, assigning the contribution to each instance for aggregating all instance-level features to a bag-level embedding. Recently, Li *et al.* [16] proposed a non-local attention aggregator that gives the contribution to each instance by the similarity between the highest-score instance and others. Shao *et al.* [23] introduced the self-attention mechanism into the MIL framework which considers the contextual and spatial information between different instances. Notably, WSI contains rich phenotypic information that reflects underlying molecular processes and disease progression. Several studies have shown phenotypic information could provide a convenient visual representation of disease aggressiveness [21,29,31]. Yao *et al.* [29] proposed a MIL framework for survival prediction that considers phenotype clusters as instances instead of patches.

Nevertheless, there are several challenges that exist in developing robust deep MIL models to learn rich representation. First, a positive WSI might contain few disease-positive patches as well as a lot of redundant instances [12,16,19,23,29], leading to the prediction failure of the models due to the weak supervision of the bag-level labels. Second, the model can easily suffer from over-fitting with limited number of training data (WSIs) [16,18] and labels.

To address these challenges, we propose a novel Clustering-Based Multi-Instance Learning (CBMIL) model, which constructs discriminative set from phenotypic clusters that highlight the significant patches of WSI. Meanwhile, a random set is constructed to augment training data for the contrastive task in our multi-task learning module. Hence, the main contributions of our work are summarized as follows:

- A novel clustering-based multi-instance learning model is proposed: it constructs discriminative set by adaptively sampling from phenotypic clusters based on the similarity between instance and phenotypic centroid.
- A mechanism for updating centroid of the phenotypic cluster is designed, which is to calculate the aggregation feature of each phenotypic cluster as the new cluster centroid in each epoch to improve the reliability of prediction.
- The contrastive learning is set as an auxiliary task of the classification task to regularize the feature aggregation process.
- CBMIL is evaluated for WSI classification on two public WSI datasets, namely: CAMELYON16 and TCGA-Lung. Great performances over these datasets and interpretability demonstrate the superiority of the proposed model compared with other state-of-the-art methods.

102   W. Wu et al.

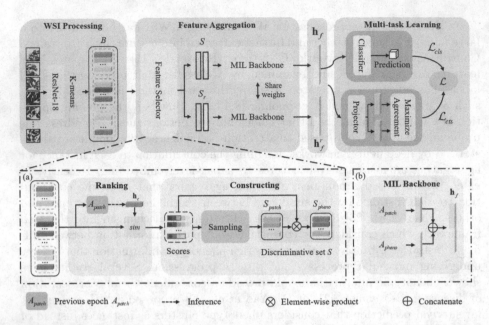

Fig. 1. The pipeline of our method. With the input feature bag, first, a feature selector constructs discriminative set S and random set S_r. Then a MIL backbone encodes the two sets to obtain high-level representations. Finally, the whole model is jointly trained by classification loss $\mathcal{L}_{cls}$ and contrastive loss $\mathcal{L}_{cts}$. (a) represents the construction of discriminative set using the selector, and (b) depicts the framework of the MIL backbone which is consisted of patch-level aggregator A_{patch} and phenotype-level aggregator A_{pheno}.

2 Method

Figure 1 depicts the overall architecture of our proposed MIL-based framework. Given an input feature bag of a WSI after clustering, two separate sets (*i.e.*, discriminative and random sets) are constructed by a feature selector, then, the selector and a MIL backbone are trained to maximize the agreement of the sets using a contrastive loss. Meanwhile, the discriminative set is involved in classification training, establishing a multi-task learning framework with the contrastive task. Specifically, in Fig. 1(a), where the construction of the discriminative set is illustrated for each training epoch. With the input feature bag, the patch-level aggregator of the previous epoch produces a sequence of centroids for each phenotypic cluster. These centroids are used to select discriminative features based on distance measurement. These discriminative features are aggregated to generate the phenotype-level features to form the discriminative set.

2.1 Clustering-Based MIL Framework

As shown in Fig. 1, a clustering-based multi-instance learning framework with multi-task learning is built for WSI classification, in which a feature selector is used to construct discriminative set fed into the MIL backbone to obtain the high-level representation (see respectively Figs. 1(a) and (b)). Then, the representation $\mathbf{h}_f$ is used to generate bag-level prediction which will be used to calculate the cross-entropy loss with the slide-level ground truth labels. Also, a small neural network projector that maps $\mathbf{h}_f$ and $\mathbf{h}_f^r$ to the latent space where contrastive loss is applied.

Let $B = \{\mathbf{B}_i\}_{i=1}^C$ denotes a bag of the clustered features of a WSI, where C is the number of clusters, $\mathbf{B}_i = \{\mathbf{x}_{i,j}\}_{j=1}^{N_i}$ is the ith phenotypic cluster that consists of patch features $\mathbf{x}_{i,j} \in \mathbb{R}^{L \times 1}$ extracted by pre-trained ResNet-18 [11] from image patches, where N_i is the number of patches of ith cluster could vary for different clusters and L is the dimension of the patch feature.

As detailed in Fig. 1(a), a discriminative set is generated by the two following processes: ranking and constructing. In the ranking phase, different non-local attention scores are assigned to patches within each cluster respectively. In a phenotypic cluster, the score of a patch is obtained based on the similarity of the patch feature to the centroid $\mathbf{h}_c$ of the cluster. The centroid is inferred from the patch-level aggregator A_{patch} of the previous epoch during training. Given a phenotypic cluster $\mathbf{B}_i = \{\mathbf{x}_{i,j}\}_{j=1}^{N_i}$, the score $r_{i,j}$ of the jth patch can be formulated as:

$$r_{i,j} = \frac{\exp(\langle \mathbf{W}_q \mathbf{h}_{c,i}, \mathbf{W}_q \mathbf{x}_{i,j} \rangle)}{\sum_{k=1}^{N_i} \exp(\langle \mathbf{W}_q \mathbf{h}_{c,i}, \mathbf{W}_q \mathbf{x}_{i,k} \rangle)}, \tag{1}$$

where $\langle \cdot, \cdot \rangle$ denotes the inner product of two vectors, and $\mathbf{W}_q$ is a weight matrix of fully-connected layer. In constructing phase, N patches are sampled with top scores from all phenotypic clusters, given by:

$$\mathbf{B}'_i = T(\mathbf{B}_i; K_i, \mathbf{r}_i), \quad K_i = \left[N_i \times \frac{N}{\sum_{k=1}^C N_k} \right], \tag{2}$$

where T is the top-k operation of choosing patches from $\mathbf{B}_i$ according to the scores $\mathbf{r}_i = \{r_{i,j}\}_{j=1}^{N_i}$. Also, K_i denotes the number of chosen patches in the ith cluster. Then, compose all the $\mathbf{B}'_i$ to get the subset of WSI $S_{patch} = \{\mathbf{x}_n\}_{n=1}^N$. Furthermore, the phenotype-level feature is aggregated by the scores within each sampled phenotypic cluster, is given by $\mathbf{x}_i^p = \sum_{j=1}^{K_i} r_{i,j} \mathbf{W}_v \mathbf{x}_{i,j}$, where $\mathbf{W}_v$ is a weight matrix used to transform $\mathbf{x}_{i,j} \in \mathbf{B}'_i$ into an information vector. The phenotype-level features are represented as $S_{pheno} = \{\mathbf{x}_i^p\}_{i=1}^C$. Finally, S_{patch} and S_{pheno} together form the discriminative set $S = \{S_{patch}, S_{pheno}\}$. Notably, the only difference between the construction of random set and discriminative set is that the features in the random set are sampled randomly and do not depend on the attention scores. As S_{patch} is sampled from the phenotypic cluster whose patches are uniformly distributed in WSI and which has the same proportion of phenotypic features as the WSI. As A_{patch} is updated during training, the selector can sample the more informative patches from each phenotypic cluster, and the model benefits from the selector as well.

Meanwhile, the network of our MIL backbone includes two feature aggregators: A_{patch} and A_{pheno}, as shown in Fig. 1(b). These two aggregators encode the constructed WSI set to patch-level and phenotype-level features which are concatenated to obtain the high-level representation of WSI. Given a WSI set $S = \{S_{patch}, S_{pheno}\}$, the fused representation $\mathbf{h}_f$ is given by:

$$\mathbf{h}_f = Cat(A_{patch}(S_{patch}), A_{pheno}(S_{pheno})), \qquad (3)$$

where Cat is a concatenation operator. With the two aggregators and concatenation operator, the MIL backbone generates a high-level representation of WSI, providing rich information for following the multi-task learning module.

2.2   Multi-task Learning

In this sub-section, a multi-task learning module is detailed, it is designed to improve the representational power of our model and mitigate over-fitting as shown in Fig. 1. Inspired by recent contrastive algorithms [2–4,9,10], we propose an auxiliary contrastive task based on our adaptive selector and MIL backbone to update our model together with the classification task.

The contrastive algorithm learns representations by maximizing agreement between differently augmented views of the same data example via a contrastive loss in the latent space [2]. The two different views of a sample are generated by a stochastic data augmentation module in previous works. Different from this, in CBMIL, a discriminative set S and a random set S_r are generated by the proposed feature selector from the same bag B, as shown in Fig. 1.

Then, the two sets from the same WSI bag as a positive pair will be transformed into two representations $\mathbf{h}_f$ and $\mathbf{h}_f^r$ by our MIL backbone. Then, the representations are mapped to vectors in latent space and $NT\text{-}Xent$ contrastive loss is applied to maximize their agreement. In addition, $\mathbf{h}_f$ is also used for the classification task, which is trained using a standard cross-entropy loss. The total loss $\mathcal{L}$ for a given mini-batch WSIs is the weighted sum of both the contrastive loss $\mathcal{L}_{cts}$ and classification loss $\mathcal{L}_{cls}$, given by $\mathcal{L} = \beta \cdot \mathcal{L}_{cts} + (1 - \beta) \cdot \mathcal{L}_{cls}$, where $\beta \in [0, 1]$ is a scalar for scaling.

2.3   Model Structure and Training Procedure

In the proposed MIL backbone, the aggregator could be an arbitrary MIL-based model that satisfies the permutation-invariant MIL formulation, such as in [12, 16,19]. We use CLAM-SB [19], a solid MIL aggregator, as our A_{patch} to aggregate the sampled features of WSI, and, A_{pheno}, a simple gated attention [12] is used to aggregate the phenotype-level features. As denoted in Eq. (2), N is a constant number that denotes the number of selected patch features. For a few WSIs with patches less than N, we will pad the bag with 0 vectors.

Stochasticity is important in contrastive learning, previous works [2,4,9,10] usually use stronger data augmentation on images. But the WSI bag is a feature-level data sample, consequently, the natural data augmentation methods are not available. To address this problem, we apply Mixup [30] based data interpolation for S_{patch} inspired by [26]. The Mixup operation is only used during the training phase. Given a mini-batch of M bags $\mathcal{B} = \{\mathbf{B}_m\}_{m=1}^{M}$ with the same constant shape, the augmented sample for a $\mathbf{B}$ is created by taking its random interpolation with another randomly chosen sample $\tilde{\mathbf{B}}$ from $\mathcal{B}$, formulated as:

$$\mathbf{B}^{+} = \lambda \cdot \mathbf{B} + (1 - \lambda) \cdot \tilde{\mathbf{B}}, \tag{4}$$

where λ is a coefficient sampled from a uniform distribution $\lambda \sim U(\alpha, 1.0)$. The value of α is usually high such as 0.9. It means that $\mathbf{B}^{+}$ is closer to $\mathbf{B}$ than $\tilde{\mathbf{B}}$, and the $\tilde{\mathbf{B}}$ could be thought of as a data noise being added.

In the inference step, we throw away the contrastive branch and the generated random set, and keep only the discriminative set for predicting the WSI label.

3   Experiments and Results

In this section, the implementation of the proposed method is detailed; also, experiments and results are reported. Our experiments are conducted on two public datasets: CAMELYONG16 [1] and the lung cancer dataset of The Cancer Genome Atlas (TCGA-Lung) [25].

3.1   Dataset and Evaluation Metrics

CAMELYON16 is a widely used public dataset for metastasis detection in breast cancer, including 270 training WSIs and 129 test WSIs. TCGA-Lung consists of two subtype projects, *i.e.*, Lung Squamous Cell Carcinoma (TGCA-LUSC) and Lung Adenocarcinoma (TCGA-LUAD), which contains 529 LUAD WSIs and 512 LUSC WSIs.

For all WSIs in both datasets, tissue segmentation of the WSI was performed by applying a combination of filters [28]. Each WSI is tiled into a series of 256×256 patches without overlap at $20\times$ magnification, where the background patches (tissue region $<35\%$) are discarded. After pre-processing, CAMELYON16 yields about 6881 patches per WSI, and TCGA-Lung yields about 11540 patches per WSI. As in [16], the feature of each patch is embedded in a 512-dimensional ($L = 512$, L is defined at the beginning of the Sect. 2.1) vector by a ResNet-18 [11] model pre-trained by [16]. Then, we adopt K-means algorithm to cluster patch features into $C = 10$ phenotypic cluster to form bag B, following [29].

Regarding CAMELYON16 dataset, the training set is done after splitting the 270 WSIs into approximately 80% training and 20% validation and tested on the official test set. For TCGA-Lung, we randomly split the data in the ratio of training:validation:test = 60:15:25. For evaluation metrics, the accuracy, Area Under Che curve (AUC) scores and F1-score are reported in Sect. 3.3 on both datasets. The average results are obtained by 4-fold cross-validation on TCGA-Lung dataset.

Table 1. Results on CAMELYON16 and TCGA-Lung, respectively.

	CAMELYON16			TCGA-Lung		
	Accuracy	AUC	F1-score	Accuracy	AUC	F1-score
MinMax [6]	0.8504	0.8757	0.7800	0.8373	0.9088	0.8396
ABMIL [12]	0.8640	0.8939	0.7988	0.8457	0.9073	0.8419
ABMIL-Gated [12]	0.8550	0.8766	0.7833	0.8468	0.9078	0.8426
SetTransformer [15]	0.7775	0.8493	0.7415	0.6758	0.7800	0.7176
DeepAttnMISL [29]	0.8791	0.9213	0.8236	0.7992	0.8744	0.79506
CLAM-SB [19]	0.8713	0.8926	0.8107	0.8687	0.9412	0.8697
CLAM-MB [19]	0.8508	0.8938	0.7866	0.8661	0.9420	0.8660
DSMIL [16]	0.8682	0.8832	0.7952	0.8597	0.9300	0.8590
CBMIL (ours)	**0.9380**	**0.9541**	**0.9184**	**0.8849**	**0.9429**	**0.8853**

3.2 Implementation Details

The number of sampled patches N is experimentally set to 1024. In the training step, we use Adam [14] optimizer with an initial learning rate of 0.0001, a cosine annealing (without warm restarts) scheme for learning rate scheduling [17], and a mini-batch size of 16. The parameter α of Mixup is set to 0.8, the temperature parameter τ defined in *NT-Xent* loss [2] is set to 1.0, and the loss scaling parameter β is set to 0.1. The classifier and projector are two Multilayer Perceptron (MLP) with one hidden layer, where the classifier calculates the prediction scores and the projector maps the representations to a 128-dimensional latent space.

3.3 Experimental Results

To demonstrate the performance of our model, we first compare our proposed model with the current state-of-the-art deep MIL models [6,12,15,16,19,29]. All the results are provided in Table 1. In CAMELYON16, only a small portion of regions in a positive slide contains tumor (roughly < 10% of the total tissue area per slide) which leads to the positive bags being highly unbalanced. CBMIL outperforms its A_{patch} CLAM-SB [19] (*i.e.*, 5% and 6% higher in accuracy and AUC) and other deep MIL-based models. In TCGA-Lung, a positive slide contains a relatively larger area of tumor region (roughly >80% of the total tissue area per slide). CBMIL also outperforms all the other methods. Overall, the results demonstrate the superiority of our CBMIL model.

In addition, to further determine the effect of the adaptive sampling mechanism and multi-task module combined with contrastive learning, we report ablation study results as shown in Table 2. This table shows the experimental results of whether our proposed model has adaptive sampling and multi-task module. Here, the A indicates whether to sample patch features based on the attention scores in the feature selector, and the M indicates whether to add contrastive learning branch in the training phase to establish multi-task learning. It could

Table 2. Effects of the adaptive sampling and multi-task module.

Method	A	M	CAMELYON16			TCGA-Lung		
			Accuracy	AUC	F1-score	Accuracy	AUC	F1-score
Random MIL			0.8915	0.9173	0.8409	0.8626	0.9344	0.8603
Adaptive MIL	✓		0.9302	0.9438	0.9032	0.8769	0.9301	0.8755
CBMIL	✓	✓	**0.9380**	**0.9541**	**0.9184**	**0.8849**	**0.9429**	**0.8853**

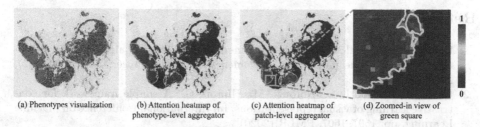

(a) Phenotypes visualization   (b) Attention heatmap of phenotype-level aggregator   (c) Attention heatmap of patch-level aggregator   (d) Zoomed-in view of green square

Fig. 2. The visualization of phenotypes and attention heatmaps: (a) is the visualization of phenotypes of a WSI from CAMELYON16 testing set, (b) and (c) are heatmaps of attention weights in aggregators. Note: for (b) and (c), attention weights are re-scaled from min-max to [0, 1] and used for patch intensities. (The details and colors are better seen by zooming on a computer screen.)

be noted that the performance of classification can be substantially improved by the adaptive sampling mechanism, and the performance can be further improved by adding the multi-task learning module.

In closing, we also show the interpretability of CBMIL as displayed in Fig. 2. The yellow curve depicts the official pixel-level annotation of the tumor region in CAMELYON16. Figure 2(a) allows the visualization of phenotypic clusters after clustering, where each color represents a cluster, and it can be noticed that the phenotypic cluster of the tumor region are very obvious, while other normal tissues are uniformly distributed throughout the WSI. It is remarkable in Fig. 2(b) that the phenotypic clusters belonging to the tumor region are given high weights. Finally, Fig. 2(c) shows a more fine-grained attention heatmap: the boundaries of which can highly overlap with the labeled regions. These visualization results demonstrate the reliable interpretability of our proposed model.

4   Conclusion

In this paper, a novel Clustering-Based Multi-Instance Learning framework (CBMIL) is proposed for weakly supervised classification of Whole Slide Image (WSI). Firstly, we design a feature selector that constructs discriminative set of WSI from phenotypic clusters by sampling patches based on centroids. The centroids are updated during training and are used to sample patches that are highly correlated with the prediction results. In addition, with the representational power of contrastive learning, we integrate contrastive learning task directly into MIL, establishing a multi-task learning framework to improve the performance of our method. Meanwhile, a Mixup operator is introduced for feature-level data augmentation. Most importantly, the proposed method outperforms the state-of-the-art MIL algorithms in terms of accuracy, AUC and F1-score over two public datasets, namely: CAMELYON16 and TCGA-Lung. Eventually, CBMIL can provide great interpretability by visualizing the attention weights in the MIL backbone.

References

1. Bejnordi, B.E., et al.: Diagnostic assessment of deep learning algorithms for detection of lymph node metastases in women with breast cancer. JAMA **318**(22), 2199–2210 (2017)
2. Chen, T., Kornblith, S., Norouzi, M., Hinton, G.: A simple framework for contrastive learning of visual representations. In: International Conference on Machine Learning, pp. 1597–1607. PMLR (2020)
3. Chen, X., Fan, H., Girshick, R., He, K.: Improved baselines with momentum contrastive learning. arXiv preprint arXiv:2003.04297 (2020)
4. Chen, X., He, K.: Exploring simple Siamese representation learning. In: Proceedings of the IEEE/CVF Conference on Computer Vision and Pattern Recognition, pp. 15750–15758 (2021)
5. Cornish, T.C., Swapp, R.E., Kaplan, K.J.: Whole-slide imaging: routine pathologic diagnosis. Adv. Anat. Pathol. **19**(3), 152–159 (2012)
6. Courtiol, P., Tramel, E.W., Sanselme, M., Wainrib, G.: Classification and disease localization in histopathology using only global labels: a weakly-supervised approach. arXiv preprint arXiv:1802.02212 (2018)
7. Dietterich, T.G., Lathrop, R.H., Lozano-Pérez, T.: Solving the multiple instance problem with axis-parallel rectangles. Artif. Intell. **89**(1–2), 31–71 (1997)
8. Feng, J., Zhou, Z.H.: Deep MIML network. In: Proceedings of the AAAI Conference on Artificial Intelligence, vol. 31 (2017)
9. Grill, J.B., et al.: Bootstrap your own latent-a new approach to self-supervised learning. Adv. Neural. Inf. Process. Syst. **33**, 21271–21284 (2020)
10. He, K., Fan, H., Wu, Y., Xie, S., Girshick, R.: Momentum contrast for unsupervised visual representation learning. In: Proceedings of the IEEE/CVF Conference on Computer Vision and Pattern Recognition, pp. 9729–9738 (2020)
11. He, K., Zhang, X., Ren, S., Sun, J.: Deep residual learning for image recognition. In: Proceedings of the IEEE Conference on Computer Vision and Pattern Recognition, pp. 770–778 (2016)
12. Ilse, M., Tomczak, J., Welling, M.: Attention-based deep multiple instance learning. In: International Conference on Machine Learning, pp. 2127–2136. PMLR (2018)
13. Kanavati, F., et al.: Weakly-supervised learning for lung carcinoma classification using deep learning. Sci. Rep. **10**(1), 1–11 (2020)
14. Kingma, D.P., Ba, J.: Adam: a method for stochastic optimization. arXiv preprint arXiv:1412.6980 (2014)
15. Lee, J., Lee, Y., Kim, J., Kosiorek, A., Choi, S., Teh, Y.W.: Set transformer: a framework for attention-based permutation-invariant neural networks. In: International Conference on Machine Learning, pp. 3744–3753. PMLR (2019)

16. Li, B., Li, Y., Eliceiri, K.W.: Dual-stream multiple instance learning network for whole slide image classification with self-supervised contrastive learning. In: Proceedings of the IEEE/CVF Conference on Computer Vision and Pattern Recognition, pp. 14318–14328 (2021)
17. Loshchilov, I., Hutter, F.: SGDR: stochastic gradient descent with warm restarts. arXiv preprint arXiv:1608.03983 (2016)
18. Lu, M.Y., Chen, R.J., Wang, J., Dillon, D., Mahmood, F.: Semi-supervised histology classification using deep multiple instance learning and contrastive predictive coding. arXiv preprint arXiv:1910.10825 (2019)
19. Lu, M.Y., Williamson, D.F., Chen, T.Y., Chen, R.J., Barbieri, M., Mahmood, F.: Data-efficient and weakly supervised computational pathology on whole-slide images. Nat. Biomed. Eng. **5**(6), 555–570 (2021)
20. Maron, O., Lozano-Pérez, T.: A framework for multiple-instance learning. Adv. Neural Inf. Process. Syst. **10** (1997)
21. Mobadersany, P., et al.: Predicting cancer outcomes from histology and genomics using convolutional networks. Proc. Natl. Acad. Sci. **115**(13), E2970–E2979 (2018)
22. Pantanowitz, L., et al.: Review of the current state of whole slide imaging in pathology. J. Pathol. Inform. **2**(1), 36 (2011)
23. Shao, Z., et al.: TransMIL: transformer based correlated multiple instance learning for whole slide image classification. Adv. Neural. Inf. Process. Syst. **34**, 2136–2147 (2021)
24. Srinidhi, C.L., Ciga, O., Martel, A.L.: Deep neural network models for computational histopathology: a survey. Med. Image Anal. **67**, 101813 (2021)
25. Tomczak, K., Czerwińska, P., Wiznerowicz, M.: The Cancer Genome Atlas (TCGA): an immeasurable source of knowledge. Wspolczesna Onkol. **2015**(1A), A68–A77 (2014)
26. Verma, V., Luong, T., Kawaguchi, K., Pham, H., Le, Q.: Towards domain-agnostic contrastive learning. In: International Conference on Machine Learning, pp. 10530–10541. PMLR (2021)
27. Wang, X., Yan, Y., Tang, P., Bai, X., Liu, W.: Revisiting multiple instance neural networks. Pattern Recogn. **74**, 15–24 (2018)
28. Yamashita, R., Long, J., Saleem, A., Rubin, D.L., Shen, J.: Deep learning predicts postsurgical recurrence of hepatocellular carcinoma from digital histopathologic images. Sci. Rep. **11**(1), 1–14 (2021)
29. Yao, J., Zhu, X., Jonnagaddala, J., Hawkins, N., Huang, J.: Whole slide images based cancer survival prediction using attention guided deep multiple instance learning networks. Med. Image Anal. **65**, 101789 (2020)
30. Zhang, H., Cisse, M., Dauphin, Y.N., Lopez-Paz, D.: mixup: beyond empirical risk minimization. arXiv preprint arXiv:1710.09412 (2017)
31. Zhu, X., Yao, J., Zhu, F., Huang, J.: WSISA: making survival prediction from whole slide histopathological images. In: Proceedings of the IEEE Conference on Computer Vision and Pattern Recognition, pp. 7234–7242 (2017)

Multi-task Learning-Driven Volume and Slice Level Contrastive Learning for 3D Medical Image Classification

Jiayuan Zhu[1], Shujun Wang[2(✉)], Jinzheng He[3], Carola-Bibiane Schönlieb[2], and Lequan Yu[4]

[1] Xi'an Jiaotong-Liverpool University, Suzhou, China
[2] DAMTP, University of Cambridge, Cambridge, UK
{sw991,cbs31}@cam.ac.uk
[3] Qilu Hospital of Shandong University, Jinan, China
[4] The University of Hong Kong, Hong Kong SAR, China

Abstract. Automatic 3D medical image classification, *e.g.*, brain tumor grading from 3D MRI images, is important in clinical practice. However, direct tumor grading from 3D MRI images is quite challenging due to the unknown tumor location and relatively small size of abnormal regions. One key point to deal with this problem is to learn more representative and distinctive features. Contrastive learning has shown its effectiveness with representative feature learning in both natural and medical image analysis tasks. However, for 3D medical images, where slices are continuous, simply performing contrastive learning at the volume-level may lead to inferior performance due to the ineffective use of spatial information and distinctive knowledge. To overcome this limitation, we present a novel contrastive learning framework from synergistic 3D and 2D perspectives for 3D medical image classification within a multi-task learning paradigm. We formulate the 3D medical image classification as a Multiple Instance Learning (MIL) problem and introduce an attention-based MIL module to integrate the 2D instance features of each slice into the 3D feature for classification. Then, we simultaneously consider volume-based and slice-based contrastive learning as the auxiliary tasks, aiming to enhance the global distinctive knowledge learning and explore the correspondence relationship among different slice clusters. We conducted experiments on two 3D MRI image classification datasets for brain tumor grading. The results demonstrate that the proposed volume- and slice-level contrastive learning scheme largely boost the main classification task by implicit network regularization during the model optimization, leading to a 10.5% average AUC improvement compared with the basic model on two datasets.

Keywords: Contrastive learning · Multi-task learning · 3D MRI image classification · Brain tumor grading

---

J. Zhu and S. Wang—Equal contribution.

1   Introduction

Automatic medical image analysis with supervised deep learning has already demonstrated promising results in various computer-assisted diagnosis tasks [23,27]. Among these tasks, 3D medical image classification (*e.g.*, brain tumor grading) is a challenging problem due to the unknown tumor location and the relatively small size of abnormal regions compared with the whole 3D volumes, especially for small medical image datasets. As the key discriminative knowledge from tumor regions is usually hidden in high-dimensional 3D input space, it is essential to balance the relationship between 2D and 3D features to improve the understanding ability of deep learning models. Currently, to exploit more discriminative information and improve the generalization capability of deep models on small datasets, a possible solution [17] is to pre-train the model on other existing labeled datasets which have a similar data distribution to the current medical dataset, followed by fine-tuning on the target dataset. However, for some rare tumour types, the annotated datasets available for transfer learning may be extremely limited, which can cause the models to be easily over-fitted.

Recently, self-supervised representation learning has demonstrated promising results with limited data. In the pre-training stage, self-supervised learning has shown its effectiveness with representative feature learning in many tasks, from both computer vision and medical image analysis [2,9,28]. Generally, it enhances the feature extraction capability of the deep models by designing proxy tasks to mine the representational properties of the data, where the data itself can be regarded as supervisory information. For example, the contrastive learning task is one of the discriminative proxy tasks [7,8,10,11,13,24], which can be used as the auxiliary task together with the main supervised task in a multi-task learning setup to improve performance for semi-supervised learning [19,25].

In this work, we aim to enhance the 3D medical image classification task under the scope of self supervision, especially contrastive learning. However, for 3D medical images, most conventional contrastive learning tasks focus on only extracting global representations or local features [1,26]. Since slices in 3D medical images are continuous, simply performing contrastive learning at the 3D volume-level may lead to inferior performance due to the ineffective use of spatial information and distinctive knowledge. To address these barriers, we employ contrastive learning from synergistic volume and slice perspectives, which is realized by two auxiliary tasks to exploit the 3D global information and 2D clustering knowledge for the 3D medical image classification. We regard the 3D medical image classification as a Multiple Instance Learning (MIL) problem and introduce a gated attention-based MIL module [16] to integrate the 2D instance features of each 2D slice into the 3D feature of the 3D volume for classification. We then adopt the volume-level contrastive learning task to enhance the global distinctive knowledge learning. While for the slice-level contrastive learning task, we divide the slices into different clusters to explore the correspondence relationship among various clusters instead of processing directly on the slices themselves, as shown in other literature [15]. Specifically, for each 3D medical image volume, we hypothesize that the slices inside the volume can

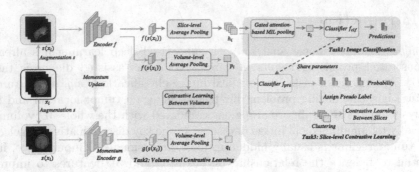

Fig. 1. Overview of our multi-task learning framework for 3D medical image classification. The given volume is first to be augmented into two views and then be fed into encoder f and momentum encoder g. The extracted features of encoder f are used for image classification (L_{clf}) via a gated attention-based MIL mechanism and also used for slice-based contrastive learning (L_{slice}). Simultaneously, they and features extracted by momentum encoder g are performed volume-level average pooling for volume-based contrastive learning (L_{volume}).

be divided into several categories according to their pseudo labels (*e.g.*, categories with/without tumor). To obtain the slice pseudo label, here we creatively propose a slice label scheme, which directly reuses the classifier on the main volume-level classification task. Then the slice-level contrastive learning can be applied to all the categories within the 3D medical volume or across multiple volumes. Experiments demonstrate that the proposed two auxiliary tasks boost the main classification task by implicit network regularization during the model optimization.

In summary, our proposed multi-task learning-driven framework contains three key modules: volume-level contrastive learning, slice-level contrastive learning, and gated attention MIL module. Through the well-designed ablations study on two downstream 3D brain MRI image classification tasks, *i.e.*, meningioma grading and glioma grading, our proposed volume-based contrastive learning improves 8.8% of AUC on average. Further equipped with slice-based contrastive learning, our method get a 9.5% AUC improvement on average compared with the baseline without any auxiliary tasks.

2   Method

In this section, we introduce our multi-task learning framework for the 3D medical image classification task. We first define the problem and give an overview of the whole framework. Then we highlight three key parts of our approach: i) gated attention-based MIL pooling for image classification; ii) volume-level contrastive learning; and iii) slice-level contrastive learning.

2.1 Problem Definition and Overview

We formulate a 3D medical image classification task under the multiple instance learning scope, where each volume represents a bag, and each slice is regarded as one instance. The overview of the proposed framework is illustrated in Fig. 1. Formally, suppose that we have N labeled 3D MRI volumes in the training dataset, $i.e.$ $\{(x_i, y_i)\}_{i=1}^{N}$. Under the contrastive learning scenario, we employ two encoders (f and g) after different augmentation methods to extract the consistent features. The encoder f and momentum encoder g have the same 3D CNN architecture but are updated differently. The parameters of encoder f is optimized along with the network training by back-propagation, while momentum encoder g is updated as a moving average of encoder parameters following

$$\theta_g = m * \theta_g + (1 - m) * \theta_f, \quad \text{where } m \in [0, 1), \tag{1}$$

where m is a momentum coefficient to control the update speed of momentum encoder g. With the moving average updating, the momentum encoder slowly updates the network parameters to avoid generating inconsistent features due to sharp changes in the encoder.

Given a 3D volume x_i and a random transformation s, the 3D volume is firstly augmented and then fed into the encoder f and momentum encoder g that outputs the discriminative feature maps $f(s(x_i))$, $g(s(x_i)) \in \mathbb{R}^{C \times D \times H \times W}$ respectively, where D, H, W are the depth, width, and height of the feature maps, and C is the number of channels. Specifically, in the encoder f, we abandon pooling operations on the z-axis to ensure that the features generated by the encoder can be partitioned into features of each slice. Then, we apply slice-level average pooling on feature maps $f(s(x_i))$ and transpose them to collect the features of each slice; thus, we get $H = \{h_i\}_{i=1}^{N}$, $h_i \in \mathbb{R}^{D \times C}$, $h_i = \{h_i^j\}_{j=1}^{D}$, $h_i^j \in \mathbb{R}^C$. It is fed into image classification task and slice-level contrastive learning task for further consideration. At the same time, we also do volume-level average pooling on $f(s(x_i))$ and $g(s(x_i))$, and get P, $Q = \{p_i\}_{i=1}^{N}$, $\{q_i\}_{i=1}^{N}$, while p_i, $q_i \in \mathbb{R}^C$, for the implementation of volume-based contrastive learning task. In the following section, we lay out three tasks sequentially.

2.2 Gated Attention-Based MIL Pooling for Image Classification

The first key part of our framework addresses the 3D medical image classification challenge under multiple instance learning - that is, how to integrate the instance (slice) features into the whole bag (volume) features. We utilize the gated attention mechanism [16] to integrate the instance features into the bag features. Compared to the conventional average pooling, it is adaptive to the characteristics of the task to achieve better performance. Let $Z = \{z_i\}_{i=1}^{N}$ be the bag features, which is obtained through gated attention pooling:

$$z_i = \sum_{j=1}^{N} a_i^j h_i^j, \tag{2}$$

where a_i is the attention weight for slice feature h_i, which is calculated based on gating mechanism [12]:

$$a_i^j = \frac{\exp\{w^T(\tanh(V(h_i^j)^T) \odot \mathrm{sigm}(U(h_i^j)^T)\}}{\sum_{k=1}^{K} \exp\{w^T(\tanh(V(h_i^k)^T)\} \odot \mathrm{sigm}(U(h_i^k)^T)\}}, \quad (3)$$

where $w \in \mathbb{R}^{L \times 1}$, $V \in \mathbb{R}^{L \times M}$, and $U \in \mathbb{R}^{L \times M}$ are the parameters, $\odot$ is an element-wise multiplication, and $\mathrm{sigm}(\cdot)$ is the sigmoid function. Then, the 3D features are fed into f_{clf} for the main image-level classification task. To optimize the main image classification task, we employ Binary Cross-Entropy loss:

$$L_{clf} = \frac{1}{N} \sum_{i=1}^{N} -(1 - y_i)\log(1 - f_{clf}(z_i)) - y_i \log f_{clf}(z_i). \quad (4)$$

2.3 Volume-Level Contrastive Learning

To explore the network understanding of the volume-level feature discrimination ability, we first introduce volume-level contrastive learning as one auxiliary task with self-supervised learning. It could enhance the distinctive global knowledge learning on the volume-level with the known volume-level annotations. Here, we employ the recently-proposed MoCo v3 [11] to implement contrastive learning among volumes. The InfoNCE [21] is employed here as the contrastive loss:

$$L_{volume} = -\log(\frac{\exp(p \cdot q^+ / \tau)}{\exp(p \cdot q^+ / \tau) + \sum_{q^-} \exp(p \cdot q^- / \tau)}), \quad (5)$$

where q^+ and p are obtained by encoder f and momentum encoder g with the same volume input x_i; Therefore, they belong to positive pairs. The set q^- consists of q's outputs from other volumes (not x_i), namely q's negative samples. τ is a temperature hyperparameter for l_2-normalization.

2.4 Slice-Level Contrastive Learning

Since the image label, i.e., volume-level annotation, is generated according to the slice-level knowledge, focusing on the slice features in 3D volumes can assist in the overall performance of the model. Contrastive learning between slices can contribute to the model to exploit more spatial-related information. However, for 3D medical image classification, the slice-level label is unknown, and only the volume-level label is available. To obtain the slice pseudo labels, here we creatively propose a slice label scheme, which directly reuses the classifier on the main image classification task. Specifically, as shown in Fig. 1, we first feed slice feature maps H into the classifier f_{pro} to assign pseudo labels to each slice according to the probability value. Here, f_{clf} and f_{pro} share network parameters. With the slice pseudo labels, we can divide the slices into two clusters by the value of generated probability. That is, the n slices with the highest probability values and n slices with the lowest probability values are separated. Then, we adopt contrastive learning between the selected features $K_i = \{k_i^j\}_{j=1}^{2n}$ with pseudo labels in each volume through minimizing a contrastive loss function. We also adopt the form of InfoNCE [21] for optimization as

$$L_{slice} = \sum_{i=1}^{N} \sum_{j=1}^{2n} -\log(\frac{\sum_{k_i^+} \exp(k_i^j \cdot k_i^+/\tau)}{\sum_{k_i^+} \exp(k_i^j \cdot k_i^+/\tau) + \sum_{k_i^-} \exp(k_i^j \cdot k_i^-/\tau)}), \quad (6)$$

where the set of k_i^+/ k_i^- consists of all the features in k_i which has the same/different label as the label of k_i^j.

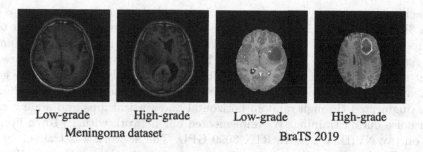

| Low-grade | High-grade | Low-grade | High-grade |

Meningoma dataset   BraTS 2019

Fig. 2. Samples of MRI images used in this paper.

3 Experiments

In this section, we provide the dataset details, data pre-processing and experimental setting, and results for our evaluation protocol.

3.1 Datasets

We used two 3D brain MRI datasets to evaluate the effectiveness of our framework. Figure 2 shows some comparison examples of the validation datasets.

Meningoma Dataset. The dataset was acquired at three hospitals in China. A total number of 507 3D brain T1c MRI volumes from 371 unique patients were collected. Specifically, a total of 356 Grade I volumes and 151 Grade II volumes are collected, respectively. The dataset is randomly divided at the patient-level to prevent the volumes from the same patient from appearing in both training and testing sets in our experiments. In detail, 321, 178, and 108 volumes are used for training, validation, and testing, respectively.

BraTS2019. BraTS 2019 training dataset [3–6,20] included 259 cases of high-grade gliomas (HGG) and 76 cases of low-grade gliomas (LGG), each with four 3D MRI modalities a native precontrast (T1), a post-contrast T1-weighted (T1Gd), a T2-weighted (T2), and a T2 Fluid Attenuated Inversion Recovery (T2-FLAIR). The volumes in the dataset were resampled to $1 \times 1 \times 1$ mm isotropic resolution and skull-stripped with the image size of $240 \times 240 \times 155$. The dataset used in this paper only contains modality T1Gd, with two types of labels: HGG and LGG. The training set, validation set, and test set contain 214, 54, and 67 volumes, respectively.

3.2 Data Pre-processing and Experimental Setup

To facilitate the training procedure, we conducted some pre-processing steps for each 3D brain volume. We first resampled the brain volumes of the dataset to a common space of $1 \times 1 \times 1$ mm. Then, we crop the black margin of the volumes and resize them to $32 \times 128 \times 128$ using bilinear interpolation. We normalize all input images to have zero mean and unit std. Additionally, to improve the generalization of the model and enhance the robustness of the network, all volumes were augmented through random horizontal and vertical flips and adding random noise during training. We implemented our network with PyTorch library [22] on two NVIDIA Geforce RTX 3090 GPU. The network was trained for 50 epochs. The batch size for our experiments was 8, and Adam optimizer [18] was used. The learning rate was set as 10^{-4}. The parameters in our network were initialized with the He initialization method [14].

Table 1. Ablation study on key components of our framework with the Meningoma dataset and BraTS 2019 dataset (AUC [%]). The best results on two datasets are highlighted in bold.

GAP	L_{Volume}	L_{Slice}	Meningoma dataset	BraTS 2019
–	–	–	84.94 ± 2.15	87.47 ± 2.08
✓	–	–	87.65 ± 1.16	94.32 ± 2.04
–	✓	–	91.92 ± 0.51	92.17 ± 1.87
–	–	✓	90.51 ± 1.65	94.64 ± 1.35
✓	✓	–	90.33 ± 1.81	96.93 ± 0.65
–	✓	✓	84.85 ± 1.83	94.53 ± 1.59
✓	–	✓	86.24 ± 4.37	92.67 ± 1.73
✓	✓	✓	$\mathbf{92.63 \pm 1.14}$	$\mathbf{97.84 \pm 0.23}$

Table 2. Comparing our framework and two-stage methods (pre-training and fine-tuning) with different components (AUC [%]). The best results on two datasets are highlighted in bold.

Pre-training	Fine-tuning	Meningoma dataset	BraTS 2019
N/A	N/A	84.94 ± 2.15	87.47 ± 2.08
N/A	Gated attention MIL	87.65 ± 1.16	94.32 ± 2.04
Volume contrastive learning	N/A	88.14 ± 2.27	89.99 ± 3.46
Volume contrastive learning	Gated attention MIL	87.94 ± 2.46	94.46 ± 3.45
Volume & Slice contrastive learning	Gated attention MIL	90.83 ± 1.32	95.39 ± 2.17
N/A	Our framework	**92.63 ± 1.14**	**97.84 ± 0.23**

3.3 Experimental Results

To evaluate the effectiveness of each proposed component, we first conduct the ablation study of the proposed framework on two datasets. The comparison results of different ablation models are listed in Table 1 with the evaluation matrix of AUC. As we can observe from Table 1, Gate attention-based pooling (GAP), volume-level, and slice-level contrastive learning consistently improve the baseline model (*i.e.*, simple 3D classification with global average pooling) by 5.5%, 7.8%, and 6.8% AUC on the average of two datasets, respectively, showing the effectiveness of our proposed each component. With the use of GAP, volume-level and slice-level contrastive learning further achieve 8.6% and 3.7% AUC improvement over the baseline, respectively. As a result, our proposed multi-task learning model equipped with the three key components achieve 10.5% mean AUC improvement over the baseline model.

Table 2 also compares the performance of the proposed framework with two-stage methods with different designed components. In the two-stage methods, we first use volume-based or slice-based contrastive learning to take the model pre-trained; subsequently in the fine-tuning stage, only the gated attention MIL is used. It is observed that our designed components also enhance the performance of the two-stage framework. Specifically, the use of gate attention-based MIL for fine-tuning led to a 5.5 % AUC improvement compared to the baseline model. Equipped with gate attention-based MIL, volume-level only as well as volume- and slice-level synergistically improve by 5.8%, and 8.0% mean AUC, respectively. Figure 3 presents ROC curves of the baseline model, two-stage method, and our method. The experiment results in Table 2, and Fig. 3 indicate that the multi-task learning framework improves the performance on both datasets, where the two-stage method with the same designed components improves only 8.0% mean AUC compared to the baseline model while our method improves 10.5% AUC averaged on two datasets.

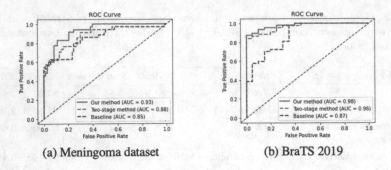

(a) Meningoma dataset   (b) BraTS 2019

Fig. 3. ROC curves for our method, the two-stage method and the baseline.

4   Conclusion

We have shown that the proposed multi-task learning method can jointly learn
3D features and 2D features by working on auxiliary tasks from 3D MRI images.
To this end, we have added two auxiliary tasks: volume-level contrastive learning
between the different views of the volumes and slice-level contrastive learning
between slice features with generated pseudo labels to extract more semantic
features, thus improving performance in the main task. Further, we apply gated
attention pooling to integrate slice features into 3D features in classification. The
ablation study on two datasets demonstrates that our proposed method leads
to substantial performance gains in the limited dataset size. In future work, we
will further investigate this approach for other downstream medical image tasks,
aiming to extend the framework to a wider variety of tasks.

Acknowledgements. The work described in this paper is supported by grants from
HKU Seed Fund for Basic Research (Project No. 202009185079 & 202111159073).
CBS acknowledges the Philip Leverhulme Prize, the EPSRC fellowship EP/V029428/1,
EPSRC grants EP/T003553/1, EP/N014588/1, Wellcome Trust 215733/Z/19/Z and
221633/Z/20/Z, Horizon 2020 No. 777826 NoMADS and the CCIMI.

References

1. Azizi, S., et al.: Big self-supervised models advance medical image classification.
   In: Proceedings of the IEEE/CVF International Conference on Computer Vision,
   pp. 3478–3488 (2021)
2. Bai, W., et al.: Self-supervised learning for cardiac MR image segmentation by
   anatomical position prediction. In: Shen, D., et al. (eds.) MICCAI 2019. LNCS,
   vol. 11765, pp. 541–549. Springer, Cham (2019). https://doi.org/10.1007/978-3-
   030-32245-8_60
3. Bakas, S. et al.: Segmentation labels and radiomic features for the pre-operative
   scans of the TCGA-GBM collection (2017). https://doi.org/10.7937/K9/TCIA.
   2017.KLXWJJ1Q

4. Bakas, S. et al.: Segmentation labels and radiomic features for the pre-operative scans of the TCGA-LGG collection (2017). https://doi.org/10.7937/K9/TCIA.2017.GJQ7R0EF

5. Bakas, S., et al.: Advancing the cancer genome atlas glioma MRI collections with expert segmentation labels and radiomic features. Sci. Data 4(1), 1–13 (2017)

6. Bakas, S., et al.: Identifying the best machine learning algorithms for brain tumor segmentation, progression assessment, and overall survival prediction in the brats challenge. arXiv preprint. arXiv:1811.02629 (2018)

7. Caron, M., Bojanowski, P., Joulin, A., Douze, M.: Deep clustering for unsupervised learning of visual features. In: Proceedings of the European conference on computer vision (ECCV), pp. 132–149 (2018)

8. Caron, M., Misra, I., Mairal, J., Goyal, P., Bojanowski, P., Joulin, A.: Unsupervised learning of visual features by contrasting cluster assignments. Adv. Neural. Inf. Process. Syst. 33, 9912–9924 (2020)

9. Chen, L., Bentley, P., Mori, K., Misawa, K., Fujiwara, M., Rueckert, D.: Self-supervised learning for medical image analysis using image context restoration. Med. Image Anal. 58, 101539 (2019)

10. Chen, T., Kornblith, S., Norouzi, M., Hinton, G.: A simple framework for contrastive learning of visual representations. In: International conference on machine learning, pp. 1597–1607. PMLR (2020)

11. Chen, X., Xie, S., He, K.: An empirical study of training self-supervised vision transformers. In: Proceedings of the IEEE/CVF International Conference on Computer Vision, pp. 9640–9649 (2021)

12. Dauphin, Y.N., Fan, A., Auli, M., Grangier, D.: Language modeling with gated convolutional networks. In: International conference on machine learning, pp. 933–941. PMLR (2017)

13. He, K., Fan, H., Wu, Y., Xie, S., Girshick, R.: Momentum contrast for unsupervised visual representation learning. In: Proceedings of the IEEE/CVF conference on computer vision and pattern recognition, pp. 9729–9738 (2020)

14. He, K., Zhang, X., Ren, S., Sun, J.: Delving deep into rectifiers: surpassing human-level performance on imagenet classification. In: Proceedings of the IEEE international conference on computer vision, pp. 1026–1034 (2015)

15. He, X., Fang, L., Tan, M., Chen, X.: Intra-and inter-slice contrastive learning for point supervised oct fluid segmentation. IEEE Trans. Image Process. 31, 1870–1881 (2022)

16. Ilse, M., Tomczak, J., Welling, M.: Attention-based deep multiple instance learning. In: International conference on machine learning, pp. 2127–2136. PMLR (2018)

17. Kamnitsas, K., et al.: Unsupervised domain adaptation in brain lesion segmentation with adversarial networks. In: Niethammer, M., et al. (eds.) IPMI 2017. LNCS, vol. 10265, pp. 597–609. Springer, Cham (2017). https://doi.org/10.1007/978-3-319-59050-9_47

18. Kingma, D.P., Ba, J.: Adam: a method for stochastic optimization. arXiv preprint. arXiv:1412.6980 (2014)

19. Koohbanani, N.A., Unnikrishnan, B., Khurram, S.A., Krishnaswamy, P., Rajpoot, N.: Self-path: self-supervision for classification of pathology images with limited annotations. IEEE Trans. Med. Imaging 40(10), 2845–2856 (2021)

20. Menze, B.H., et al.: The multimodal brain tumor image segmentation benchmark (brats). IEEE Trans. Med. Imaging 34(10), 1993–2024 (2014)

21. Oord, A.V.D., Li, Y., Vinyals, O.: Representation learning with contrastive predictive coding. arXiv preprint. arXiv:1807.03748 (2018)

22. Paszke, A., et al.: Automatic differentiation in pytorch (2017)
23. Wang, W., Chen, C., Ding, M., Yu, H., Zha, S., Li, J.: TransBTS: multimodal brain tumor segmentation using transformer. In: de Bruijne, M., et al. (eds.) MICCAI 2021. LNCS, vol. 12901, pp. 109–119. Springer, Cham (2021). https://doi.org/10.1007/978-3-030-87193-2_11
24. Zbontar, J., Jing, L., Misra, I., LeCun, Y., Deny, S.: Barlow twins: self-supervised learning via redundancy reduction. In: International Conference on Machine Learning, pp. 12310–12320. PMLR (2021)
25. Zhai, X., Oliver, A., Kolesnikov, A., Beyer, L.: S4l: self-supervised semi-supervised learning. In: Proceedings of the IEEE/CVF International Conference on Computer Vision, pp. 1476–1485 (2019)
26. Zhou, H.Y., Lu, C., Yang, S., Han, X., Yu, Y.: Preservational learning improves self-supervised medical image models by reconstructing diverse contexts. In: Proceedings of the IEEE/CVF International Conference on Computer Vision, pp. 3499–3509 (2021)
27. Zhou, Z., et al.: Models genesis: generic autodidactic models for 3d medical image analysis. In: Shen, D., et al. (eds.) MICCAI 2019. LNCS, vol. 11767, pp. 384–393. Springer, Cham (2019). https://doi.org/10.1007/978-3-030-32251-9_42
28. Zhuang, X., Li, Y., Hu, Y., Ma, K., Yang, Y., Zheng, Y.: Self-supervised feature learning for 3D medical images by playing a Rubik's cube. In: Shen, D., et al. (eds.) MICCAI 2019. LNCS, vol. 11767, pp. 420–428. Springer, Cham (2019). https://doi.org/10.1007/978-3-030-32251-9_46

Light Annotation Fine Segmentation: Histology Image Segmentation Based on VGG Fusion with Global Normalisation CAM

Yilong Li[1,2], Yaqi Wang[1]([✉]), Le Dong[3], Juan Ye[4], Linyan Wang[4], Ruiquan Ge[5], Huiyu Zhou[6], and Qianni Zhang[2]([✉])

[1] Communication University of Zhejiang, Hangzhou, China
wangyaqi@cuz.edu.cn
[2] Queen Mary University of London, London, UK
qianni.zhang@qmul.ac.uk
[3] University of Electronic Science and Technology of China, Chengdu, China
[4] The Second Affiliated Hospital of Zhejiang University, Hangzhou, China
[5] Hangzhou Dianzi University, Hangzhou, China
[6] University of Leicester, Leicester, UK

Abstract. Deep learning has been widely used to segment tumour regions in stained histopathology images. However, precise annotations are expensive and labour-consuming. To reduce the manual annotation workload, we propose a light annotation-based fine-level segmentation approach for histology images based on a VGG-based Fusion network with Global Normalisation CAM. The experts are only required to provide a rough segmentation annotation on the images, and then accurate fine-level segmentation boundaries can be produced using this method. To validate the proposed approach, three datasets with rough and fine quality segmentation annotation are built. The fine quality labels are used only as ground truth in evaluation. The VFGN-CAM method includes three main components: an annotation enhancement to boost boundary accuracy and model generalisability; a VGG Fusion module that integrates multi-scale tumour features; and a Global Normalisation CAM module that combines local and global gradient information of tumour regions. Our VGG fusion and Global Normalisation CAM outperform the existing methods with a Dice of 84.188%. The final improvement for our proposed methods against the original rough labels is around 22.8%.

Keywords: Segmentation · Tumor · Annotation improvement

1 Introduction

Cancer is the most deadly illness in the world due to it capability to generate distant metastases. Digital pathology scanners can provide whole slide image (WSI)

W. Qin et al. (Eds.): CMMCA 2022, LNCS 13574, pp. 121–130, 2022.
https://doi.org/10.1007/978-3-031-17266-3_12

with a very high resolution (e.g. 80000 × 150000). Stained WSIs are the gold standard for diagnosing cancer and predicting tumour reoccurrence and other potential deterioration. However, manual tumour segmentation is expensive and time consuming for pathologists. Therefore, the automatic segmentation method is essential for efficient and accurate tumour classification on WSIs.

Several challenges exist in labelling tumour regions. Compared with carefully hand-drawn boundaries that describe exactly the tissue structures, pathologists tend to mark tumour parts with rough smooth curves in practice, which will save substantial time for marking. The rough markings are informative but could be misleading in model training to some extent since these boundaries include inevitable errors. In addition, Intrinsic variance in the tumour, the variances between patients, and technical variances generated in slicing, staining, and scanning cause inaccurate manual tumour segmentation annotation.

To relieve the dependence on segmentation annotation, many methods have been proposed for weakly supervised segmentation (WSS) purposed including image-level [2], scribble-based [4], point-based [10] and iterative based methods [11]. Class activation mapping(CAM) [14] with global average pooling (GAP) is a simple yet effective technique for weakly-supervised segmentation. Wang et al. propose consistency regularization on predicted CAMs from various transformed images to provide self-supervision [12]. Durand et al. jointly aim at aligning image regions for gaining spatial invariance and learning strongly localized features [1]. Similar to CAM, adversarial erasing is an efficient way to represent objects partly according to the peak responses of classes [3,7]. Recently, Multi-branch WSS methods are proposed to segment objects more preciously such as complex attention modules [5], cross-image mining [8] and siamese networks [12]. Most of the existing methods are designed by combining a series of modules including training classifiers, visualizing activation maps and re-training segmentation networks.

Inspired by the efficacy of WSS methods, to reduce the dependency on accurate tumour annotations and minimise pathologists' workload of marking on WSIs, we build two kinds of annotations including fine quality labels(F-label), and poor quality labels(P-label). The purpose of this work is to exploit a large amount of P-labels for training and use a few F-labels for testing.

In this work, we propose a VGG-based fusion network with global normalization CAM (VFGN-CAM). Our contributions are threefold. (1) We refine the P-labels based on k-means clustering and soft label. This annotation refinement process ensures the annotation accuracy of tumour boundaries and enhances the subsequent model generalizability. (2) A VGG-based fusion module (VF-Net) is proposed based on VGG16. Multi-scale features are fused together for patch-based tumour classification. (3) A global normalization CAM (GN-CAM) module is presented to integrate gradient information both in the global whole image and local patches, to acquire the position features in distinguishing the tumour and background.

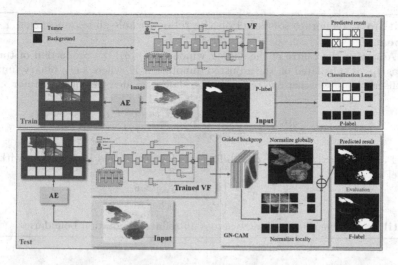

Fig. 1. The structure of VFGN-CAM. VF-Net are trained with the data pre-processed by annotation enhancement (AE), GN-CAM are used in test stage to acquire more accurate result.

2   Methods

The overall framework is shown in Fig. 1. The rough annotation is first processed in the annotation enhancement (AE) module which employs the k-means clustering algorithm to improve the annotation for network training with soft labels. Then we propose a VGG-based fusion classification network based on VGG16 to exploit multi-scale features for fine-grained patch-based classification. After network training, the information of the last convolution layer of the network is extracted and calculated by a GN-CAM which combines the normal CAM result and a global normalization CAM result by specific weights. At last, the output heat-map for each patch is embedded into the whole slide image and then goes through a convolutional CRFs-based noise eliminator (NE) to smooth the boundary of the generated annotation and eliminate the noise.

2.1   Annotation Enhancement

To reduce the inaccuracy of the rough annotations, we propose an annotation enhancement (AE) module based on k-means clustering, and a soft label modifier to refine cancer annotations, especially in tumor marginal regions. The rough annotation Y_0 marked by experts delineates non-tumour regions from tumour regions. K-means clustering cluster together pixels with similar features together to create label Y_1 that is a refined version of the original tumour boundaries in Y_0. The intersection point set $\hat{Y} = Y_1 \cap Y_0$ is considered as the refined ground truth. In addition, to ensure a highly efficient model training and boost the model generalizability, patch-based soft labels [13] are generated by a sliding window with the size of (512,512) as shown in Algorithm 1.

Algorithm 1. Generate soft label on the refined whole slide annotation $\hat{Y}$

1: **repeat**

2:    Assuming the centre of the sliding window is (p, q), the proportion of tumour areas $f_{p,q}$ is calculated on the adjusted annotation Y, where $\mathbb{I}$ is a binary function to discriminate whether one region in the sliding window belongs to tumor.

$$f_{p,q} = \sum_{i=p-256}^{p+256} \sum_{j=p-256}^{p+256} \frac{\mathbb{I}(tumor, i, j)}{512^2}$$

3:    Patches X extracted by the sliding window are stored and marked with soft label Y

$$Y = \begin{cases} 1 - \sigma, & f_{p,q} > \theta \\ \sigma, & others \end{cases}$$

4: **until** moving the sliding window across all refined annotation boundaries.

As several errors still exist in tumour boundaries and especially in isolated tiny tumour regions, the errors will be propagated if we directly train models according to pixel-wise refined annotations Y. In this case, we design a patch-based classification model with GN-CAM supervised by soft labels, to reduce the error effects around tumour boundaries.

2.2   VF Classification Network

Convolution-based design is capable of inferring accurate local features (texture, boundary and greyscale) with few features. VGG is a universal backbone for image feature extraction, which has been widely applied to classification, detection and segmentation tasks for medical images [6].

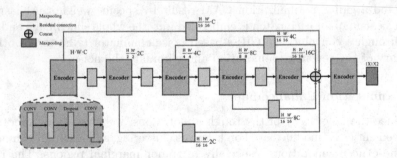

Fig. 2. The network structure of VF.

Convolution-based design is capable of inferring accurate local features (texture, boundary and greyscale) with few features. Inspired by the effectiveness and lightweight of VGG [6], we apply VGG 16 as our base model for tumour classification, increasing a series of residual connections among convolutions and design a multi-scale fusion module to ensure accurate classification of tiny tumours. Figure 2 illustrates the detailed framework of our VF method. Each block contains three convolution layers with residual connections to ensure the stability of network back-propagation. One dropout layer is inserted after the second convolution layer to increase the network generalization. In addition, a multi-scale feature fusion module is presented to fuse feature maps generated by all Max-pooling layers. All feature maps are resampled to the same size as the feature map from the final Max-pooling. These features are concatenated together and pass through a convolution layer.

2.3   Global Normalised Class Activation Mapping

Global Normalised Class Activation Mapping (GN-CAM) is a new way of combining feature maps using the gradient signal. Inspired by assigning an importance factor to each neuron by the gradient of G-CAM, this paper proposes a global normalized CAM that extracts the guided gradient features R^l flowing out the last convolution layer. The lth and $(l+1)^{th}$ layers are the last two layers of the VF. Denote the ith feature map of the $(l+1)^{th}$ layer as f_i^{l+1}, the ith gradient map of the $l+1$ layer as R_i^{l+1} and the output map is f^{out}. The gradient map of $l+1$ layer is calculated by

$$f_i^{l+1} = relu\left(f_i^l\right) = max\left(f_i^l, 0\right), \tag{1}$$

$$R_i^{l+1} = \partial f^{out}/\partial f_i^{l+1}. \tag{2}$$

The guided gradient map of the l layer R^l is calculated by:

$$R_i^l = \left(f_i^l > 0\right) \cdot \left(R_i^{l+1} > 0\right) \cdot R_i^{l+1}. \tag{3}$$

All guided gradient maps R from the same WSI are stored in a queue Q_1. Then we normalise each map in Q_1 with the global mean and standard deviation. These processed maps R' are stored in a new queue Q_2. Also, assuming (w, h) is the spatial position of a gradient map R, every pixel $R_{i,w,h}^l$, $w \in W, h \in H$ is normalized locally and recalculated by:

$$\mu_i^l = (\sum_{w=1}^{W}\sum_{h=1}^{H} R_{i,w,h}^l)/WH \tag{4}$$

$$s_i^l = \sqrt{[\sum_{w=1}^{W}\sum_{h=1}^{H}\left(R_{i,w,h}^l - \mu_i^l\right)^2]/(WH)^2} \tag{5}$$

$$R_i^{l''} = (R_i^l - \mu_i^l)/s_i^l \tag{6}$$

The locally normalised maps R'' from the same WSI are stored in a queue Q_3. Two normalised gradient maps R_i' and R_i'' from Q_2 and Q_3 are added together by order. The final segmentation results M is calculated by

$$M_i = (R_i' + R_i'')/2. \tag{7}$$

2.4 Noise Eliminator

After model training and the CAM process, the generated masks are more accurate, but some noise remains. This is because the tumour regions are calculated in the region of 512 × 512 pixels, so the predicted boundary is very sharp. Also, the isolated tumour cells and fine details in boundaries are often not considered in human manual labelling. Thus, to resemble manual segmentation, a post-processing step using convolutional CRFs [9] is developed to ensure the segmentation boundary is medically relevant. The output of convolutional CRFs has more smooth boundaries and less noise, especially inside the tumour region.

3 Experiment and Result

3.1 Data Introduction and Training Details

We train and evaluate our framework on three tumour datasets including basal cell cancer (BCC), squamous papilloma (SP), and seborrheic keratosis cancer (SKC) datasets. All three datasets are skin cancer data. The common challenge of a skin cancer dataset is that the boundary of the tumour region is difficult to identify. So the rough annotations on this kind of dataset will further influence the performance of the segmentation network. In the training process, to reduce the requirement for memory and accelerate the training process, we cut all the whole slide tumour images into patches the size of (512,512). The Adam optimizer is used with a learning rate of 0.0001 and a step learning scheduler with step size = 60 and $\gamma = 0.95$. The loss function is cross entropy.

3.2 Evaluation and Results

There are five widely used measurement parameters used in this evaluation: sensitivity, specificity, accuracy, IOU and dice coefficient. IOU and dice coefficient are widely used to comprehensively evaluate the segmentation performance of the target network. First, we discuss the performance of annotation enhancement and our proposed VF network. Table 1 demonstrates the evaluation about the annotation enhancement (AE) and network. All network results with annotation enhancement have a better performance against the same network without annotation enhancement. It is reasonable to believe that our proposed AE has

Table 1. Network results with or without annotation enhancement.

	Sensitivity(%)	Specificity(%)	Accuracy(%)	IOU(%)	Dice(%)
VGG	70.480	97.735	93.997	61.077	75.358
VGG-AE	74.592	**97.972**	94.819	64.602	78.032
VF	75.267	97.869	94.915	64.522	77.766
VF-AE (ours)	**80.947**	97.703	**95.813**	**68.538**	**81.090**

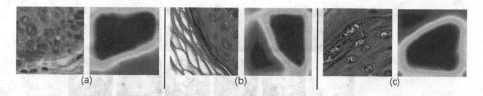

Fig. 3. Output patches of GN-CAM for three dataset: (a) basal cell cancer (BCC); (b) squamous papilloma (SP); (c) seborrheic keratosis cancer (SKC).

a non-negligible effect in a weakly trained segmentation task, especially in the poor quality annotation situation. Also, our proposed VF network has an average of 3% improvement against the VGG network. Which proves the VFGN-CAM structure is more suitable for this work.

Table 2. Results of different CAM based on annotation enhancement.

		Sensitivity(%)	Specificity(%)	Accuracy(%)	IOU(%)	Dice(%)
VGG	CAM	69.386	**98.212**	94.010	61.574	75.697
	GN-CAM	79.799	97.732	95.628	67.629	80.367
VF	CAM	78.890	97.790	95.597	67.569	80.316
	GN-CAM	**83.003**	97.615	**96.029**	**69.507**	**81.865**

CAM aims to extract the information in the convolution layer to explain the results of network training. In this work, information is extracted from the last convolution layer of the trained network. We propose a global normalization CAM in tumour segmentation task, the result of this CAM of patches is shown in Fig. 3. We explore the differences between two kinds of CAM, the segmentation result are shown in Table 2. Using annotation enhancement or not, our proposed GN-CAM achieve better performance in all parameter against the normal CAM.

Table 3. Results of noise eliminator under annotation enhancement and GN-CAM.

	Sensitivity(%)	Specificity(%)	Accuracy(%)	IOU(%)	Dice(%)
VGG	79.799	97.732	95.628	67.629	80.367
VGG-NE	81.961	**98.106**	96.187	70.837	82.626
VF	83.003	97.615	96.029	69.507	81.865
VF-NE (ours)	**85.461**	98.000	**96.600**	**72.963**	**84.188**

As shown in Table 3, average segmentation improvements on three datasets are around 3% with the proposed noise eliminator, regardless of whether we use VGG or VF. Table 4 shows the comparison of the P-label and the predicted

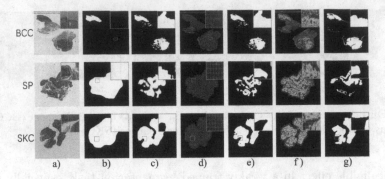

Fig. 4. Some examples of CAM output heat map and tumor segmentation results: (a) original image; (b) Poor quality label (P-label); (c) annotation for VGG; (d) heat map for VGG; (e) annotation for VF; (f) heat map for VF; (g) Fine quality label (F-label).

segmentation evaluated against the F-label. All the machine learning results are generated under the annotation enhancement and noise eliminator by GN-CAM. The generated mask presents a great improvement against the P-label. Among all the results, our proposed VF and GN-CAM with annotation enhancement and noise eliminator achieve the best result. Figure 4 shows the result of CAM heat map output and final annotations after noise eliminator. Compare to poor quality labels (P-label), our method generates more accurate and detailed boundaries. The proposed VF methods lead to an improvement of 22.846% in Dice coefficient against the P-label, which proves the success of the VFGN-CAM framework. Also, the heat map generated by GN-CAM shows a significant visual correlation to the tumour area, meaning that the output segmentation can be used for medical assessment tasks which have roughly annotated training sets.

Table 4. Segmentation results compared to original P-label by F-label as ground truth.

		Sensitivity(%)	Specificity(%)	Accuracy(%)	IOU(%)	Dice(%)
	P-label	57.601	97.898	87.972	45.048	61.342
CAM	VGG	70.821	**98.907**	94.473	65.037	78.600
	VF	80.569	98.329	96.162	71.001	82.759
GN-CAM	VGG	81.961	98.106	96.187	70.837	82.626
	VF	**85.461**	98.000	**96.600**	**72.963**	**84.188**

4   Conclusion

In this paper, we explore a new patch-based tumour segmentation method supervised by rough annotations called VFGN-CAM. More specifically, an annotation enhancement is presented to progressively refine the annotations, which ensures accuracy in tumour boundary shape. A VF net is used to classify the patches.

We also propose a GN-CAM to integrate global and local gradient information of tumour regions. Experiments on three tumour datasets show the effectiveness and superiority of our model. In future, more weakly supervised work will be proposed based on our P-labels and our method.

Acknowledgement. This work was supported by the Zhejiang Provincial Natural Science Foundation of China (No. LY21F020017, 2022C03043), National Natural Science Foundation of China (No. 61702146).

References

1. Durand, T., Mordan, T., Thome, N., Cord, M.: Wildcat: weakly supervised learning of deep convnets for image classification, pointwise localization and segmentation. In: Proceedings of the IEEE Conference on Computer Vision and Pattern Recognition, pp. 642–651 (2017)
2. Huang, Z., Wang, X., Wang, J., Liu, W., Wang, J.: Weakly-supervised semantic segmentation network with deep seeded region growing. In: Proceedings of the IEEE Conference on Computer Vision and Pattern Recognition, pp. 7014–7023 (2018)
3. Kweon, H., Yoon, S.H., Kim, H., Park, D., Yoon, K.J.: Unlocking the potential of ordinary classifier: class-specific adversarial erasing framework for weakly supervised semantic segmentation. In: Proceedings of the IEEE/CVF International Conference on Computer Vision, pp. 6994–7003 (2021)
4. Luo, X., et al.: Scribble-supervised medical image segmentation via dual-branch network and dynamically mixed pseudo labels supervision. arXiv preprint. arXiv:2203.02106 (2022)
5. Qin, J., Wu, J., Xiao, X., Li, L., Wang, X.: Activation modulation and recalibration scheme for weakly supervised semantic segmentation. arXiv preprint. arXiv:2112.08996 (2021)
6. Simonyan, K., Zisserman, A.: Very deep convolutional networks for large-scale visual recognition. arXiv preprint. arXiv:1409.1556 (2014)
7. Stammes, E., Runia, T.F., Hofmann, M., Ghafoorian, M.: Find it if you can: end-to-end adversarial erasing for weakly-supervised semantic segmentation. In: Thirteenth International Conference on Digital Image Processing (ICDIP 2021), vol. 11878, p. 1187825. International Society for Optics and Photonics (2021)
8. Sun, G., Wang, W., Dai, J., Van Gool, L.: Mining cross-image semantics for weakly supervised semantic segmentation. In: Vedaldi, A., Bischof, H., Brox, T., Frahm, J.-M. (eds.) ECCV 2020. LNCS, vol. 12347, pp. 347–365. Springer, Cham (2020). https://doi.org/10.1007/978-3-030-58536-5_21
9. Teichmann, M.T., Cipolla, R.: Convolutional CRFs for semantic segmentation. arXiv preprint. arXiv:1805.04777 (2018)
10. Tian, K., et al.: Weakly-supervised nucleus segmentation based on point annotations: a coarse-to-fine self-stimulated learning strategy. In: Martel, A.L., et al. (eds.) MICCAI 2020. LNCS, vol. 12265, pp. 299–308. Springer, Cham (2020). https://doi.org/10.1007/978-3-030-59722-1_29
11. Wang, X., You, S., Li, X., Ma, H.: Weakly-supervised semantic segmentation by iteratively mining common object features. In: Proceedings of the IEEE Conference on Computer Vision and Pattern Recognition, pp. 1354–1362 (2018)

12. Wang, Y., Zhang, J., Kan, M., Shan, S., Chen, X.: Self-supervised equivariant attention mechanism for weakly supervised semantic segmentation. In: Proceedings of the IEEE/CVF Conference on Computer Vision and Pattern Recognition, pp. 12275–12284 (2020)
13. Yuan, L., Tay, F.E., Li, G., Wang, T., Feng, J.: Revisiting knowledge distillation via label smoothing regularization. In: Proceedings of the IEEE/CVF Conference on Computer Vision and Pattern Recognition, pp. 3903–3911 (2020)
14. Zhou, B., Khosla, A., Lapedriza, A., Oliva, A., Torralba, A.: Learning deep features for discriminative localization. In: Proceedings of the IEEE conference on computer vision and pattern recognition. pp. 2921–2929 (2016)

Tubular Structure-Aware Convolutional Neural Networks for Organ at Risks Segmentation in Cervical Cancer Radiotherapy

Xinran Wu[1,5], Ming Cui[2], Yuhua Gao[2], Deyu Sun[2], He Ma[5], Erlei Zhang[6], Yaoqin Xie[1], Nazar Zaki[3,4], and Wenjian Qin[1(✉)]

[1] Shenzhen Institute of Advanced Technology,
Chinese Academy of Sciences, Shenzhen, China
huavhuahua@163.com
[2] Department of Gynecological Radiotherapy, Cancer Hospital of China Medical University, Shenyang, Liaoning, People's Republic of China
[3] Department of Computer Science and Software Engineering,
College of Information Technology, UAEU, Al Ain, UAE
[4] Big Data Analytics Center (BIDAC), UAEU, Al Ain, UAE
[5] College of Medicine and Biological Information Engineering,
Northeastern University, Shenyang, China
[6] Northwest A and F University, Yangling, Shannxi, China

Abstract. Cervical cancer is the most frequent cancer type among women worldwide and radiotherapy is the major clinical treatment. Organs in the radiation field are called Organ at Risks (OARs), which are prone to irreversible damage during radiotherapy. Therefore, accurate delineation of OARs is a critical step in ensuring radiotherapy dosimetry accuracy. However, currently existing deep learning-based cervical cancer OARs segmentation methods do not make full advantage of anatomical information. In this paper, we develop a novel tubular structure-aware deep convolutional network method integrating the tubular anatomical morphological features into a model for colon, small intestine and rectum in cervical cancer OARs. Firstly, a tubular filter based on variable annular Gaussian kernel and gradient detection was used to produce the tubular feature map. Secondary, tubular feature map concatenated with original image was input into the nnU-Net network for anatomical morphological information learning. Finally, we evaluated our proposed method on the clinical collection datasets with brachytherapy. Compared to the baseline model and state-of-the-art model, DSC and Recall were improved and the relative volume error (RVE) was reduced for the OARs with a tubular shape.

Keywords: Cervical cancer organ at risk · Anatomical information · Tubular filter

© The Author(s), under exclusive license to Springer Nature Switzerland AG 2022
W. Qin et al. (Eds.): CMMCA 2022, LNCS 13574, pp. 131–140, 2022.
https://doi.org/10.1007/978-3-031-17266-3_13

1 Introduction

According to the research of the World Health Organisation (WHO), cervical cancer is one of the top four most frequently diagnosed malignant tumors worldwide as well as the second most common cancer in women [1]. The treatment of cervical cancer mainly relies on radiotherapy [2], and the organs near the tumor in the irradiation fields are susceptible to irreversible damage caused by radiation, and these organs are called organs at risks (OARs). The delineation of OARs in CT images is indispensable in the course of radiotherapy, which is time-consuming and labor-intensive manually. Therefore, an automatic segmentation method of cervical cancer OARs is urgently demanded.

In recent years, deep learning has been widely used in organ segmentation and has become an important method for automatic segmentation, however, there are few outstanding results in the segmentation of colon, small intestine, and rectum in cervical cancer OARs. Existing cervical cancer OARs segmentation methods mostly rely on simple deep learning models without incorporating anatomical information [3,4]. Some works use filters to extract features from images to achieve organ segmentation, such as the work of Merveille et al. [5] and Krissian et al. [6]. But these methods are based on the assumption that the target is a smooth, relatively straight tubular structure with approximately circular in cross section, which are not suitable for intestinal segmentation in CT images. Orellana et al. [7] designed a tubular filter for the colon and achieved satisfactory segmentation results, but their method was not fully automated and still required some manual operations by experts. In this paper, considering the approximate tubular anatomical features of colon, small intestine and rectum in cervical cancer OARs, a Tubular Structure-Aware Convolutional Neural Networks(TSACNN) for OARs segmentation in cervical cancer is proposed. Based on the nnU-Net network [8], a tubular filter based on variable annular Gaussian kernel and gradient detection [7] is integrated into nnU-Net to enhance the tubular morphological features of OARs in the image. The extracted tubular feature map is merged with the original image and input into the nnU-Net network, and the network will output the results of multi-target segmentation of cervical cancer OARs.

Our work has two main contributions: First, we develop the TSACNN model for the segmentation of cervical cancer OARs, and incorporating anatomical features into the deep learning model enhances the performance and interpretability of the method. Before inputting the data into the nnU-Net model, a tubular filter is used to enhance the tubular feature of the image. Then the extracted tubular feature map is input to the nnU-Net network merged with the original image and the network will output the result of multi-target segmentation. Secondly, we evaluate our proposed method on a clinical cervical CT image dataset consisting of 163 3D CT volumes with delineation of organs at risk, and experiments on the datasets demonstrate the effectiveness of our method.

2 Method

In this paper, the proposed TSACNN is based on the nnUNet network for multi-object segmentation and a tubular filter is introduced to perform the segmentation of cervical cancer OARs. Figure 1 shows the architecture of our TSACNN model.

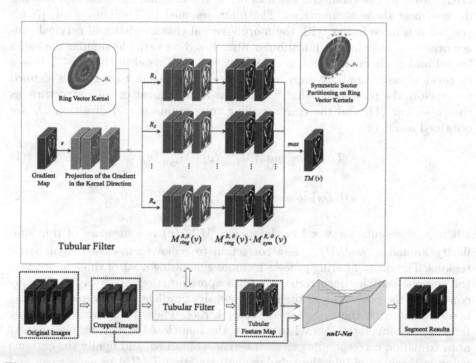

Fig. 1. The architecture of the TSACNN model for cervical cancer OARs segmentation.

As shown in Fig. 1, the whole process includes image preprocessing, tubular filtering, and deep learning model training to obtain the final segmentation result.

Since there are blank background areas in the original CT image, which will interfere with the model and cause high computational cost, the data is first preprocessed. In the data preprocessing stage, the redundant areas are removed from the original image, and the original image is first binarized to get the preliminary tissue region and background region. However, there are still artifacts formed by auxiliary devices such as CT scanning bed. Secondly, the binary image is then subjected to an open morphological operation to separate the tissue from the non-tissue part. By calculating the vertex coordinates of the minimum bounding box of the tissue region in the binary image and cropping it from the original image. Finally, the blank background area in the original image can be removed.

In order to obtain the anatomical structure information of colon, small intestine, and rectum in cervical cancer OARs, the tubular feature map of the image is obtained by the tubular filtering module. Most existing works on tubular filters make idealized assumptions on tubular targets, such as that the cross-section of the tube is a regular circle [6], and the gradient along the tube wall is zero [9]. However, the segmentation targets with tubular features in the cervical cancer OARs, that is, the geometric features of the colon, small intestine and rectum, do not meet these assumptions. The filter designed by Orellana et al. [7] for the colon is more in line with the morphological characteristics of cervical cancer organs at risk, in which a tubular filter based on variable annular Gaussian kernel and gradient detection is employed. For each voxel in the 3D CT image of cervical cancer radiotherapy, degree of approximation to a tube is defined to measure the probability that it conforms to a segment of tube structure, as shown in Eq. (1), and the corresponding tube radius R and direction D_θ are obtained as Eq. (2).

$$TM(v) = \max_{R,\theta}(M_{ring}^{R,\theta}(v) \cdot M_{sym}^{R,\theta}(v)) \tag{1}$$

$$\{R, D_\theta\} = \underset{R,\theta}{argmax}(M_{ring}^{R,\theta}(v) \cdot M_{sym}^{R,\theta}(v)) \tag{2}$$

where v represents the voxel in the image. $M_{ring}^{R,\theta}(v)$ is a measure of ring similarity around v, and $M_{sym}^{R,\theta}$ is a correction to remove non-closed asymmetric regions. The entire filtering process includes generating a set of ring-shaped vector kernels, and the uncorrected ring-like approximation $M_{ring}^{R,\theta}$ is obtained by convolving the projection of the image intensity gradient in the direction of the kernel plane with the ring-shaped vector kernel. Then, by using a symmetric fan-shaped convolution kernel to remove the non-closed structure, the annular approximation corresponding to each kernel is obtained, and finally the extreme value is taken to obtain the tubular approximation $TM(v)$ of each voxel and the corresponding tube radius R and direction D_θ, and the tubular feature map corresponding to the image is obtained. The specific detailed calculation process can be found in [7]. An example of the tubular feature map and the ground truth labels of the cervical cancer OARs region given by experts on the cross-section are shown in the Fig 2.

It can be seen that the tubular structure in the filtered image is more obvious, and the contrast between most intestinal regions and surrounding tissues has been improved. This is more conducive to extracting tubular information during model training, thereby assisting in the segmentation of the colon, intestine and rectum.

After filtering, since the tubular feature map can enhance the tubular structure features of cervical cancer OARs and the original CT image still contain rich information that the tubular feature map does not have, the extracted tubular feature map and the original image are input into the network together. Deep learning segmentation models have been widely used in medical image segmentation, and nnU-Net [8], as a mature and stable model, has achieved good results in

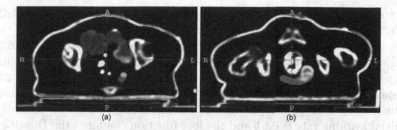

Fig. 2. An example of the tubular feature map and the ground truth labels of the cervical cancer OARs region given by experts on the cross-section. The binary part is the tubular feature map, and the colored part superimposed above is the ground truth position of each OAR (red is the bladder, green is the colon, blue is the intestine, and yellow is the colon). (a) slice example containing bladder, colon, intestine; (b) slice example containing rectum.

many tasks. And some existing work has confirmed that multi-organ segmentation results tend to be better than single-organ segmentation [10,11], therefore, this paper chooses nnU-Net as the basic network model. The nnU-Net network will output the result of multi-object segmentation.

In the model testing and evaluation stage, the test set images are first pre-processed and filtered in the same way as the training set, and the test set images and filtering results are input into the nnU-Net model in the same way as the training set. After the model inference, the segmentation results are compared with the ground truth labels delineated by doctors to calculate relevant indicators to evaluate the model effect.

3 Experiments

In this section, we first introduce the dataset and experimental environment, and then evaluate the performance of our proposed TSACNN method through comparison with existing methods and ablation experiments.

3.1 Dataset and Experimental Environment

The cervical cancer data set used in this paper comes from the data collected during 163 internal radiation treatments of 38 cervical cancer patients provided by Liaoning Cancer Hospital under ethical institutional review board approval. The images taken for each treatment are a set of data which includes the original 3D CT images of the patient's pelvis and nearby areas, as well as the labels of the four major cervical cancer OARs containing bladder, colon, small intestine and rectum, manually delineated by the doctor layer by layer. The images in the dataset are all 3D CT volumes. The original image size is (512, 512, ~100), that is, the original size of the image section on the sagittal plane of the human body is 512 voxels × 512 voxels, and the original size in the vertical axis direction is about 100 voxels. The voxel spacing of the images in the dataset is 0.976563 mm × 0.976563 mm × 3 mm.

The specific data set allocation in this experiment is: a total of 163 sets of data, each set of data includes the original image and OARs segmentation labels, and the data is randomly allocated into 90 sets of training sets, 23 sets of validation sets, and 50 sets of test sets. Set the nnU-Net network mode to 3d_fullres in addition to the parameters automatically configured by nnU-Net. The initial learning rate is 0.01 and the loss function combines the Dice loss and the cross-entropy loss function. The activation function uses leaky ReLU and the optimizer uses SGD with Nesterov momentum. For training parameters, the number of epochs is set to 1000, and the mini-batches are set to 250. The patch size is the maximum size that computing resources can support, and is automatically calculated by nnU-Net.

The experiments in this paper are carried out on the ubuntu system with 8 Core CPU, 64GB memory, and Quadro_RTX_8000 GPU.

3.2   Comparing to Existing Methods

In order to evaluate the effectiveness of the model, DSC, Recall and relative volume error RVE are calculated on the segmentation results of the model and the labels given by experts. The calculation formula is as Eq. (3)–(5).

$$DSC = \frac{2TP}{FP + 2TP + FN} \tag{3}$$

$$Recall = \frac{TP}{TP + FN} \tag{4}$$

$$RVE = \frac{abs(|R_a| + |R_b|)}{|R_b|} \tag{5}$$

where TP represents the number of correctly predicted positive voxels, FP represents the number of wrongly predicted voxels in the negative example and FN represents the number of incorrectly predicted voxels in the positive example. $|R_a|$ and $|R_b|$ represent the volume of the predicted result and the true label, respectively. It is important to emphasize that RVE is a value that indicates superior performance with lower values.

Due to the lack of public data sets with ground truth in the field of cervical cancer OARs segmentation, existing works basically use data collected by themselves or data provided by cooperative hospitals, so the results indicators of different works are not comparable with each other. Therefore, in order to better evaluate the method in this paper, some existing classic or state-of-the-art 3D image segmentation models are used to conduct comparative experiments on the dataset used in this paper. After removing the background as the method in this paper, the images are segmented by different methods. The average values of DSC obtained on test set are shown in Table 1 and the average values of Recall and RVE are shown in Fig. 3.

Table 1. The average value of DSC obtained by each method on the test set.

Methods	Bladder	Colon	Intestine	Rectum
TSACNN(ours)	**0.9033**	**0.5751**	**0.5664**	**0.6200**
nnU-Net [8]	0.9007	0.5456	0.5413	0.5946
V-Net [12]	0.6705	0.2601	0.3506	0.2913
GLIA-Net [13]	0.8068	0.2646	0.3682	0.4689
Skip DenseNet 3D [14]	0.6070	0.3538	0.3937	0.4506

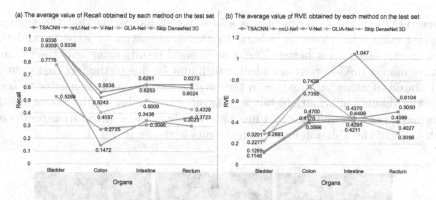

Fig. 3. The average value of Recall and RVE obtained by each method on the test set.

The bold font in the table represents the best performance of the item. It can be seen that the TSACNN method used in this paper is better than other methods in terms of the segmentation results of cervical cancer OARs on the dataset used in this paper. In Table 1, the performance of our methods on this task is better on DSC. In Fig. 3, our method generally outperforms other methods on Recall and RVE. Although the RVE of rectum is slightly higher than the other two methods, TSACNN has significantly higher DSC and Recall, so our method is better overall.

3.3   Ablation Experiments

Table 2 shows the results of the ablation experiments. The segmentation results of TSACNN are compared with the nnU-Net network model without filters and the model that only put the tubular feature map into the nnU-Net network.
The bold font in the table represents the best performance of the item. As shown in Table 2, the DSC and Recall of the colon, intestine and rectum are improved after adding the tubular filter to the nnU-Net model. It shows that the tubular filter effectively improves the overall segmentation effect of the segmentation model without adding the tubular filter, and reduces the situation of missing segmentation. The relative volume error RVE of the segmentation of all cervical

Table 2. The results of the ablation experiments.

Organs	TSACNN			nnU-Net			only Tubular Feature		
	DSC	Recall	RVE	DSC	Recall	RVE	DSC	Recall	RVE
Bladder	**0.9033**	0.9300	**0.1146**	0.9007	**0.9336**	0.1265	0.8460	0.8724	0.1746
Colon	**0.5751**	**0.5638**	**0.3996**	0.5456	0.5243	0.4170	0.5048	0.4842	0.5040
Intestine	**0.5654**	**0.6261**	**0.4211**	0.5413	0.6253	0.4370	0.4788	0.5399	0.6856
Rectum	**0.6200**	**0.6273**	**0.4089**	0.5946	0.6014	0.5050	0.5613	0.5845	0.5648

cancer OARs in the experiment was also reduced, indicating that the segmentation results were closer in volume to the labels given by experts, and the segmentation effect was better. The experimental results using only the tubular feature map in nnU-Net are not ideal, which may be due to the fact that there is still abundant other information in the original CT image. It is narrow to keep only the tubular information, but it is feasible to emphasize the characteristic information of the tubular structure. An example of the segmentation results on each OAR of the ablation experiment results is shown in Fig. 4.

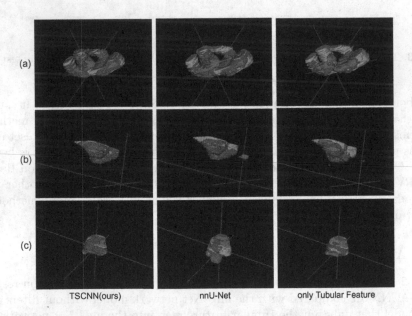

Fig. 4. An example of the segmentation results on each OAR of the ablation experiment results. Blue represents the region where the prediction is consistent with the ground truth, red represents the redundancy of the prediction, and green represents the lack of the prediction. (a)Colon; (b)Intestine; (c)Rectum.

In Fig. 4, it can be seen that the TSACNN method is still different from the ground truth, however the omission of the segmentation results of the TSACNN method has been significantly alleviated compared with the other two methods, and more complete OARs regions have been marked. Reducing the omission of segmentation results is very important in OARs segmentation work, and our TSACNN model shows better performance in this regard. At the same time, the redundancy of the segmentation results is also improved when using TSACNN, which is especially obvious when segmenting Intestine and Rectum compared to nn-UNet.

4   Conclusion and Discussion

In this paper, considering that the colon, small intestine, and rectum have approximately tubular anatomical characteristics in cervical cancer OARs, a tubular filter is introduced to the nnU-Net network to improve the performance of the segmentation network for intestinal segmentation. Compared with existing methods on the our data set, and the method in this paper has achieved the best results. Based on the anatomical and morphological characteristics of different organs, this provides an idea for future organ segmentation work, more suitable filters for different kinds of organ can be designed to assist the segmentation method. Our proposed method is not complicated and can be flexibly embedded in various deep learning models. Moreover, it can also be used for other OARs segmentation tasks, such as liver cancer, bladder cancer, prostate cancer, and rectal cancer, in which also involve OARs with similar tubular structure. In future work, the idea of morphological structure-aware deep learning can also be used in attention mechanisms or cascaded networks. Besides, the experimental results are still affected by the problem that experts ignore regions far from the tumor when delineating OARs during cervical cancer brachytherapy. Therefore, there is still room for further improvement of the segmentation effect in future work.

Acknowledgements. This work was supported by the National Natural Science Foundation of China (No. 61901463 and U20A20373), and the Shenzhen Science and Technology Program of China grant JCYJ20200109115420720, and the Youth Innovation Promotion Association CAS(2022365). The authors would like to acknowledge support from the Big Data Analytics Center (BIDAC) at the United Arab Emirates University (UAEU).

References

1. Wild, C., Weiderpass, E., Stewart, BW.: World Cancer Report: Cancer Research for Cancer Prevention. IARC Press (2020)
2. Vu, M., Yu, J., Awolude, O.A., Chuang, L.: Cervical cancer worldwide. Curr. Probl. Cancer **42**(5), 457–465 (2018)

3. Men, K., Dai, J., Li, Y.: Automatic segmentation of the clinical target volume and organs at risk in the planning CT for rectal cancer using deep dilated convolutional neural networks. Med. Phys. **44**(12), 6377–6389 (2017)

4. Liu, Z., et al.: Segmentation of organs-at-risk in cervical cancer CT images with a convolutional neural network. Phys. Med. **69**, 184–191 (2020)

5. Merveille, O., Talbot, H., Najman, L., Passat, N.: Tubular structure filtering by ranking orientation responses of path operators. In: Fleet, D., Pajdla, T., Schiele, B., Tuytelaars, T. (eds.) ECCV 2014. LNCS, vol. 8690, pp. 203–218. Springer, Cham (2014). https://doi.org/10.1007/978-3-319-10605-2_14

6. Krissian, K., Malandain, G., Ayache, N., Vaillant, R., Trousset, Y.: Model-based detection of tubular structures in 3d images. Comput. Vis. Image Underst. **80**(2), 130–171 (2000)

7. Orellana, B., Monclús, E., Brunet, P., Navazo, I., Bendezú, Á., Azpiroz, F.: A scalable approach to T2-MRI colon segmentation. Med. Image Anal. **63**, 101697 (2020)

8. Isensee, F., Jaeger, P.F., Kohl, S.A., Petersen, J., Maier-Hein, K.H.: nnU-Net: a self-configuring method for deep learning-based biomedical image segmentation. Nat. Methods **18**(2), 203–211 (2021)

9. Bauer, C., Bischof, H., Beichel, R.: Segmentation of airways based on gradient vector flow. In: International Workshop on Pulmonary Image Analysis, Medical Image Computing and Computer Assisted Intervention, pp. 191–201. Citeseer (2009)

10. Guo, D., et al.: Organ at risk segmentation for head and neck cancer using stratified learning and neural architecture search. In: Proceedings of the IEEE/CVF Conference on Computer Vision and Pattern Recognition, pp. 4223–4232 (2020)

11. Roth, H.R., et al.: Hierarchical 3D fully convolutional networks for multi-organ segmentation. arXiv preprint. arXiv:1704.06382 (2017)

12. Milletari, F., Navab, N., Ahmadi, S.A.: V-net: fully convolutional neural networks for volumetric medical image segmentation. In: 2016 Fourth International Conference on 3D Vision (3DV), pp. 565–571. IEEE (2016)

13. Bo, Z.-H., et al.: Toward human intervention-free clinical diagnosis of intracranial aneurysm via deep neural network. Patterns **2**(2), 100197 (2021)

14. Bui, T.D., Shin, J., Moon, T.: Skip-connected 3D DenseNet for volumetric infant brain MRI segmentation. Biomed. Signal Proc. Control **54**, 101613 (2019)

Automatic Computer-Aided Histopathologic Segmentation for Nasopharyngeal Carcinoma Using Transformer Framework

Songhui Diao[1,2], Luyu Tang[1,2], Jiahui He[1], Hanqing Zhao[1], Weiren Luo[3], Yaoqin Xie[1], and Wenjian Qin[1(✉)]

[1] Shenzhen Institute of Advanced Technology, Chinese Academy of Sciences, Shenzhen 518055, China
wj.qin@siat.ac.cn
[2] Shenzhen College of Advanced Technology, University of Chinese Academy of Science, Shenzhen 518055, China
[3] Shenzhen Third People's Hospital, Shenzhen 518112, China

Abstract. The segmentation of the histopathological whole slide images (WSIs) of nasopharyngeal carcinoma (NPC) plays an essential role in the diagnosis, grading and even prognosis analysis. Due to the huge size of pathological images and the fact that NPC often occurs in the middle and advanced stages, it is still challenging to generate accurate segmentation results automatically. Although many convolutional neural network (CNN) methods had achieved good segmentation performance in many types of images, however, the encoding of global context is insufficient, and it is prone to misjudge the adjacent regions. Meanwhile, the area of NPC pathological image is dense, which means that the image with a tiny size may fall into one category. To overcome this limitation, we apply a transformer-based framework on NPC pathological images that is designed for extracting and encoding global context information. To validate and compare the transformer framework with various CNN-based methods, experiments have been conducted on the clinical dataset collection of NPC. The transformer framework outperformed the state-of-the-art pure CNN-based methods in AUC and recall. Especially, our framework achieved 2.5%–3.5% higher DSC in 5X images and 2.1%–3.2% higher DSC in 10X images than other methods.

Keywords: Nasopharyngeal carcinoma · Histopathological whole slide images · Transformer · Segmentation

1 Introduction

Nasopharyngeal carcinoma (NPC), a subgroup of head and neck cancers, can be categorized into non-keratinized nasopharyngeal carcinoma or keratinized NPC [1]. According to GLOBOCAN 2018 data statistics [2], there were about 129,079 new cases and

---

S. Diao and L. Tang—Contributed equally to this work.

72,987 deaths in NPC. Typically, a pathological diagnosis is the gold standard of a cancer diagnosis. Through the pathological diagnosis of the upper respiratory tract tissue of the patient [3], it is very beneficial to confirm the tumor grade results and follow-up treatment plan [4]. Unfortunately, most patients were diagnosed only in the middle or advanced stage of cancer due to the particularity of NPC [5]. In the routine clinical diagnosis, assessing the diagnosis and its subtypes of NPC requires visual inspections by experienced pathologists under a microscope [6], which is prone to inter-observer and intra-observer variability. Hence, segmenting the cancer region of NPC is the basic procedure for subsequent diagnosis and treatment by quantitatively calculation [7]. Nevertheless, it is labor-intensive and time-consuming to observe the whole slide pathologic images for pathologists, which is highly dependent on expert knowledge. Therefore, there is a strong demand for an automatic method for quickly and accurately detecting cancer regions of NPC.

Recently, numerous automatic methods for cancer region segmentation have been proposed in WSIs. For instance, Feng *et al.* [8] modified the VGG network to segment colorectal cancer based on Unet [9]. Sun *et al.* [10] proposed multi-scale embedding networks for segmenting cancerous regions of various sizes, in which they have integrated Atrous Spatial Pyramid Pooling module and encoder-decoder based semantic-level embedding networks. Diao *et al.* [11] introduced a weakly supervised framework based on a multiscale attention convolutional neural network (MSAN-CNN) to detect the hepatocellular carcinoma cancer regions. Similarly, some classical CNN based on deep learning has been successfully applied to segmentation tasks [12–14]. There are few studies on the automatic processing of NPC pathological images. In our previous work [6], we proposed a framework to diagnose the cropped patches for NPC pathological images, in which the pixels in the patches are all the same categories, such as cancer or lymph. However, this classification work only focuses more on the internal structure and information in a patch, which is not suitable for the segmentation network considering the adjacent information. Moreover, the CNN-based methods have limitations in modelling explicit long-range relationships due to the characteristic of convolution operation i.e. locality. These structures usually produce weak performance, especially in terms of texture, shape, and size, showing different target structures in different patients, which is crucial for diagnosing hematoxylin-eosin-stained pathological images.

To solve the problems mentioned above, transformer network [15], a network considering the relationship between global information and sub-information, is an alternative method for extracting context features in visual tasks [16, 17], including pathological image classification [18] and segmentation [19]. Moreover, as shown in Fig. 1, the staining difference between the NPC and adjacent regions is minor. Because the NPC patients are in the middle and advanced stage, many areas in the tissue samples are cancer. If the cropped patches were fed into a network for training, the region of a patch is almost a cancer category. Which is highly unfavourable for our subsequent segmentation tasks. We need a large field of view to learn more abundant features. Therefore, a transformer with a solid ability to model global contexts is an excellent choice for NPC pathological image segmentation.

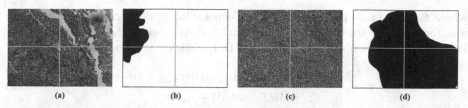

(a)   (b)   (c)   (d)

Fig. 1. Examples of the training pair images from the WSI. (a) and (c) are images to be segmented. (b) and (d) are corresponding masks. The orange line is the reference cropping line. If the patch was cropped in this line, almost all areas are one category.

In this work, we adopted a transformer-based framework, the TransUNet network [16], to extract the NPC pathological image feature and segment the cancer region. In this method, the transformer is used for feature extraction, and the skip connection of Unet is fused for decoding to calculate the final segmentation output. In the coding phase, the encoder with the transformer can capture global information, including long-distance relationships and dependencies. To raise the convergence rate of model training, we have performed the transfer learning with pretrained on ImageNet. To our best knowledge, we are the first to apply a transformer-based framework for NPC pathological image segmentation. Meanwhile, we carried out experiments at two different magnifications and achieved state-of-the-art performance compared to CNN-based methods. Moreover, the transformer-based framework had a higher recall and less risk of missed diagnosis.

2 Method

For a given pathological image of NPC $x \in \mathbb{R}^{H \times W \times C}$, where the $H \times W$ is the spatial resolution, and the number of channels C is 3, our goal is to compute the image segmentation mask. Unlike the CNN-based method, following the TransUNet, our framework adopted the transformer network with self-attention mechanisms into the encoder. We first cut the image into the target size due to the pathological image with gigapixel. Then, we apply the transformer to encode the feature representation of the image to be segmented. And the obtained features are decoded to obtain the final segmented output. Finally, we joint all the segmented outputs to get the target mask.

2.1 Encoded Stage

The input x is first tokenized as a sequence by reshaping, which consists of all flattened 2D patches. The formula is $x_p \in \mathbb{R}^{N \times (P^2 \times C)}$, where the patch size is $P \times P$, and $N = HW/P^2$ is the number of patches. This effective sequence length N finally passed into the transformer. Then, the x_p is linearly projected into a potential D-dimensional embedding space by full connection. In subsequent calculations, D is taken as the vector length, where $D = p^2 \times C$. In order to contain the spatial information of different patches, the position embeddings are added to the patch embeddings as follows:

$$z_0 = [x_p^1 E; x_p^2 E; ...; x_p^N E] + E_{pos}, E \in \mathbb{R}^{(N+1) \times D} \tag{1}$$

where E is the linear projection, and E_{pos} denotes the position embedding. For each transformer, there are two blocks, Multihead Self-Attention (MSA) and Multi-Layer Perceptron (MLP) with L layers, which can be formulated as follows:

$$z'_l = MSA(LN(z'_{l-1})) + z'_{l-1},$$
$$z_l = MLP(LN(z'_l)) + z'_l, \tag{2}$$

where LN (•) means the layer normalization and z_l is the encoded image representation. Figure 2 includes the overall segmentation process. Finally, the output after multiple transformer layers is reshaped and convoluted to encode the input x.

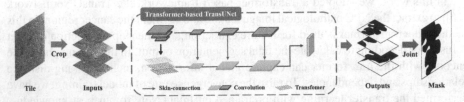

Fig. 2. Overview of the transformer-based method used in this study for segmentation of NPC.

Instead of using a transformer to extract features directly from the original image, we first use multiple convolutions to obtain 1×1 patches as the inputs of patch embeddings. Choosing this strategy can directly obtain some high-resolution feature maps, which is beneficial for the subsequent decoding stages.

2.2 Decoded Stage

After obtaining the coding feature representation $z_L = \mathbb{R}^{HW/P^2 \times D}$, a common solution is to directly upsample it to the segmentation mask of the target output. However, there is a problem that the size obtained by the decoder module is much smaller than the target mask size, which will lead to the loss of numerous low-level information and details. This is even more important for pathological images owning to some required texture and structure information. Therefore, we follow the strategy of [16], and the encoded features with different resolutions are fused to the decoder by skip connections to compute the final target mask.

Specifically, after reshaping the encoded features, the highest-level decoded features F_D^0 with shape $(D, \frac{H}{P}, \frac{W}{P})$ are obtained by performing the convolution and activation operation (CA) consisting of a 3×3 convolution layer and a ReLU activation function. Meanwhile, features of different resolutions are obtained by convolution operation in the encoded stage. Then, as shown in Fig. 2, the D_0 is $2\times$ upsampling, and concatenated with the features of the corresponding resolution from the encoded stage to decode the next-level F_D^1. After three times of this processing strategy and upsampling, the output segmentation mask is finally calculated by a CA and a segmentation head, which consists of 3×3 convolution layer and a bilinear upsampling.

3 Experiment and Result

3.1 Dataset and Evaluation

NPC2020 Dataset [6]. The NPC dataset consists of 277 cases classified as non-keratinizing carcinoma according to WHO histologic classification. The pathological images were collected from 2004 to 2018 and ranged in age from 18 to 71 years at the Department of Pathology, the People's Hospital of Gaozhou and Shenzhen Third People's Hospital. They were scanned at 40X magnification and annotated by two pathologists with at least fifteen years of clinical experience. We randomly split all WSIs into 470 training slides datasets and 312 testing slides datasets. Slides belonging to the same case will only appear in the training dataset or test dataset. In actual training, training images and corresponding labels were cropped and resized from the WSIs in 10X magnification with size 512×512 and 5X magnification with size 256×256. Meanwhile, our experiment carried out a 5-fold cross-validation in the training dataset.

Evaluation Metrics. We used four standard and conventional metrics to evaluate the performance of our framework: Dice Similarity Coefficient (DSC), Intersection over Union (IOU), Precision (PRE) and Recall (REC). The experimental results are expressed in this form: mean $\pm$ std.

3.2 Implementation Details

The routine data augmentations were applied in our experiments, such as rotation and flipping. The encoder modules, consisting of ResNet-50 [20] and ViT [17], were pre-trained on ImageNet [21]. The number of skip connections was 3. The input resolutions were set as 256 or 512 according to the magnification. Patch size of the transformer input was set as 16. For the training, the optimizer was SGD with 0.01 learning rate, $1e-4$ weight decay and 0.9 momentum. The batch size and epoch were 32 and 150. All models were implemented using PyTorch (version 1.9.0), and all training processes were trained on the two NVIDIA RTX A6000 GPU in Linux (version is 4.4.0-116-generic).

In the inference stage, patches were cropped in an overlapping strategy with 25% of the width and 30% of the height. For the pixel category of the over overlap part, we used a one vote affirmative method to select the category, that is, if there was any positive, the category of mask pixel was positive.

Table 1. Comparison on NPC2020 datasets (mean ± std).

Model	DSC	IOU	PRE	REC
	5X magnification			
Unet	0.779 ± 0.005	0.663 ± 0.007	0.823 ± 0.003	0.774 ± 0.008
FPN	0.773 ± 0.008	0.655 ± 0.010	0.817 ± 0.006	0.772 ± 0.010
Linknet	0.775 ± 0.008	0.658 ± 0.009	0.826 ± 0.005	0.769 ± 0.015
PSPNet	0.769 ± 0.003	0.650 ± 0.004	0.812 ± 0.005	0.771 ± 0.004
PAN	0.773 ± 0.009	0.655 ± 0.011	0.816 ± 0.006	0.776 ± 0.012
DeeplabV3++	0.779 ± 0.004	0.664 ± 0.005	0.823 ± 0.004	0.776 ± 0.009
Our	**0.804 ± 0.005**	**0.684 ± 0.007**	**0.829 ± 0.001**	**0.800 ± 0.008**
	10X magnification			
Unet	0.801 ± 0.003	0.696 ± 0.005	0.823 ± 0.009	0.822 ± 0.013
FPN	0.812 ± 0.001	0.707 ± 0.003	0.820 ± 0.009	0.840 ± 0.010
Linknet	0.807 ± 0.005	0.701 ± 0.006	0.832 ± 0.004	0.826 ± 0.011
PSPNet	0.803 ± 0.003	0.695 ± 0.003	0.820 ± 0.006	0.834 ± 0.004
PAN	0.808 ± 0.002	0.704 ± 0.004	0.833 ± 0.005	0.828 ± 0.001
DeeplabV3++	0.811 ± 0.002	0.707 ± 0.002	0.828 ± 0.003	0.838 ± 0.007
Our	**0.833 ± 0.001**	**0.730 ± 0.002**	**0.847 ± 0.005**	**0.856 ± 0.007**

3.3 Results

Main experiments were conducted on NPC2020 dataset by comparing the transformer network i.e. TransUNet with six previous state-of-the-art methods: 1) Unet [11]; 2) FPN [22]; 3) Linknet [13]; 4) PSPNet [14]; 5) PAN [12]; 6) DeeplabV3++ [23].

Quantitative Analysis. Our segmentation results can be found in Table 1. In general, the proposed framework outperformed all pure convolutional networks presented here by a significant margin in all metrics on the NPC2020 dataset, regardless of which magnification the results come from. Meanwhile, it can be found that the results of experiments with 10X images had achieved a better performance than that with 5X images, which indicated that the high magnification NPC images with more abundant information were beneficial to transformer framework learning. Specifically, our framework achieved 2.5%–3.5% higher DSC in 5X images and 2.1%–3.2% higher DSC in 10X images than the compared methods, which highlighted the robustness of our framework. Moreover, the performance improvement was focused on the REC metric, which illustrated that the transformer could effectively reduce false negatives. According to the above quantitative results, the transformer-based framework was more suitable for NPC pathological image segmentation of cancer regions than the pure CNN-based network.

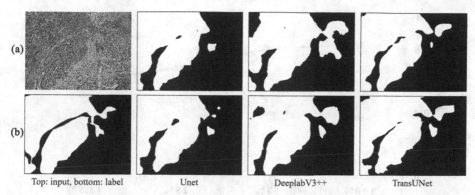

Top: input, bottom: label   Unet   DeeplabV3++   TransUNet

Fig. 3. Qualitative test results of different methods. The first column showed one sample randomly selected from a WSI. The remaining columns were the results of different models. (a) The results based on 5X. (b) The results based on 10X (black means normal area and white mean tumor area in the ground-truth and prediction results).

Qualitative Analysis. To verify the performance of the transformer-based framework, as shown in Fig. 3, qualitative comparison results were provided in the NPC2020 dataset. In order to better show the qualitative results, the baseline model Unet and DeeplabV3++ with the best overall performance were selected as the comparison models. It can be seen that: 1) for some minor areas (e.g., in the upper left corner), our framework will not ignore that compared with the pure CNN-based network. 2) Based on the 10X image, the boundary region could be segmented more smoothly, especially the transformer-based network, which showed that it has stronger power to encode global context. 3) Regardless of the image segmentation based on which magnification, the missed diagnosis of the transformer-based model was lower (i.e., the white area would not be divided into the black area). Moreover, the proposed pathological segmentation framework could perform finer segmentation to retain complete shape information for the regions with two holes in the upper right. This verifies again that the performance transformer-based framework was excellent.

Discussion. The training image of the transformer-based framework considering global context information needs the image with a larger field of view to learn more abundant features. Experiments based on TransUNet with different sizes of images were designed to verify this hypothesis. Based on the resolution of the original image (100%), we conducted experiments on 50%, 25% and 12.5% proportion size images, as shown in Fig. 4. It can be seen that the image based on 50% size still has a relatively large field of view, so the corresponding result decline is not prominent. However, the results based on 25% and 12.5% size images were collapsing due to the sharp reduction of their field of view, that is, all regions were one category, so it was not easy to get global context information. These experimental results also verified our conjecture that images with a larger field of vision were more beneficial to the transformer-based framework and conformed to the diagnosis process of pathologists.

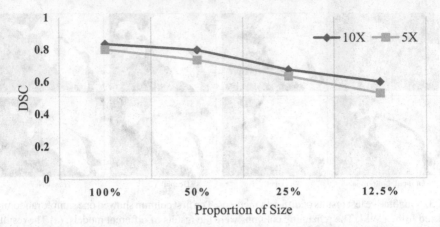

Fig. 4. Line chart of results with different proportions sizes images. The blue and red lines indicate the measurement using the input magnifications with 10X and 5X, respectively.

4   Conclusion

This paper first investigated the validity of a transformer-based framework on tumor segmentation of NPC pathological images. We use a TransUNet framework that encodes the global context by transforming the pathological image features of different regions as sequences, which can fully utilize low-level features. Comparison experiments demonstrated that the transformer framework achieves superior performance to various CNN-based methods in the NPC2020 testing dataset. Moreover, from the quantitative perspective in segmentation results, the transformer framework considers more global information and achieves a better segmentation mask for NPC pathological images.

Acknowledgements. This work was supported by the National Natural Science Foundation of China (No. 61901463 and U20A20373), and the Shenzhen Science and Technology Program of China grant JCYJ20200109115420720, and the Youth Innovation Promotion Association CAS (2022365).

References

1. Thompson, L.D.: Update on nasopharyngeal carcinoma. Head Neck Pathol. **1**, 81–86 (2007)
2. Bray, F., Ferlay, J., Soerjomataram, I., Siegel, R.L., Torre, L.A., Jemal, A.: Global cancer statistics 2018: GLOBOCAN estimates of incidence and mortality worldwide for 36 cancers in 185 countries. CA: Cancer J. Clin. **68**, 394–424 (2018)
3. Lee, H.M., Okuda, K.S., González, F.E., Patel, V.: Current perspectives on nasopharyngeal carcinoma. In: Rhim, J.S., Dritschilo, A., Kremer, R. (eds.) Human Cell Transformation. AEMB, vol. 1164, pp. 11–34. Springer, Cham (2019). https://doi.org/10.1007/978-3-030-22254-3_2
4. Liu, Y., et al.: Tumour heterogeneity and intercellular networks of nasopharyngeal carcinoma at single cell resolution. Nat. Commun. **12**, 1–18 (2021)
5. Wei, W.I., Sham, J.S.: Nasopharyngeal carcinoma. Lancet **365**, 2041–2054 (2005)

6. Diao, S., et al.: Computer-aided pathologic diagnosis of nasopharyngeal carcinoma based on deep learning. Am. J. Pathol. **190**, 1691–1700 (2020)
7. Wei, K.R., Xu, Y., Liu, J., Zhang, W.-J., Liang, Z.-H.: Histopathological classification of nasopharyngeal carcinoma. Asian Pac. J. Cancer Prev. **12**, 1141–1147 (2011)
8. Feng, R., Liu, X., Chen, J., Chen, D.Z., Gao, H., Wu, J.: A deep learning approach for colonoscopy pathology WSI analysis: accurate segmentation and classification. IEEE J. Biomed. Health Inform. **25**, 3700–3708 (2021)
9. Ronneberger, O., Fischer, P., Brox, T.: U-net: convolutional networks for biomedical image segmentation. In: Navab, N., Hornegger, J., Wells, W.M., Frangi, A.F. (eds.) MICCAI 2015. LNCS, vol. 9351, pp. 234–241. Springer, Cham (2015). https://doi.org/10.1007/978-3-319-24574-4_28
10. Sun, M., Zhang, G., Dang, H., Qi, X., Zhou, X., Chang, Q.: Accurate gastric cancer segmentation in digital pathology images using deformable convolution and multi-scale embedding networks. IEEE Access **7**, 75530–75541 (2019)
11. Diao, S., et al.: Weakly supervised framework for cancer region detection of hepatocellular carcinoma in whole-slide pathologic images based on multiscale attention convolutional neural network. Am. J. Pathol. **192**, 553–563 (2022)
12. Li, H., Xiong, P., An, J., Wang, L.: Pyramid attention network for semantic segmentation. arXiv preprint arXiv:1805.10180 (2018)
13. Chaurasia, A., Culurciello, E.: Linknet: exploiting encoder representations for efficient semantic segmentation. In: 2017 IEEE Visual Communications and Image Processing (VCIP), pp. 1–4. IEEE (2017)
14. Zhao, H., Shi, J., Qi, X., Wang, X., Jia, J.: Pyramid scene parsing network. In: Proceedings of the IEEE Conference on Computer Vision and Pattern Recognition, pp. 2881–2890 (2017)
15. Vaswani, A., et al.: Attention is all you need. Adv. Neural Inf. Process. Syst. **30** (2017)
16. Chen, J., et al.: Transunet: transformers make strong encoders for medical image segmentation. arXiv preprint arXiv:2102.04306 (2021)
17. Dosovitskiy, A., et al.: An image is worth 16x16 words: transformers for image recognition at scale. arXiv preprint arXiv:2010.11929 (2020)
18. Shao, Z., Bian, H., Chen, Y., Wang, Y., Zhang, J., Ji, X.: Transmil: transformer based correlated multiple instance learning for whole slide image classification. Adv. Neural Inf. Process. Syst. **34**, 2136–2147 (2021)
19. Nguyen, C., Asad, Z., Deng, R., Huo, Y.: Evaluating transformer-based semantic segmentation networks for pathological image segmentation. In: Medical Imaging 2022: Image Processing, pp. 942–947. SPIE (2022)
20. He, K., Zhang, X., Ren, S., Sun, J.: Deep residual learning for image recognition. In: Proceedings of the IEEE Conference on Computer Vision and Pattern Recognition, pp. 770–778. IEEE (2016)
21. Deng, J., Dong, W., Socher, R., Li, L.-J., Li, K., Fei-Fei, L.: Imagenet: a large-scale hierarchical image database. In: 2009 IEEE Conference on Computer Vision and Pattern Recognition, pp. 248–255. IEEE (2009)
22. Kirillov, A., He, K., Girshick, R., Dollár, P.: A unified architecture for instance and semantic segmentation (2017)
23. Chen, L.-C., Zhu, Y., Papandreou, G., Schroff, F., Adam, H.: Encoder-decoder with atrous separable convolution for semantic image segmentation. In: Ferrari, V., Hebert, M., Sminchisescu, C., Weiss, Y. (eds.) ECCV 2018. LNCS, vol. 11211, pp. 833–851. Springer, Cham (2018). https://doi.org/10.1007/978-3-030-01234-2_49

Accurate Breast Tumor Identification Using Computational Ultrasound Image Features

Yongqing Li[1] and Wei Zhao[1,2(✉)]

[1] School of Physics, Beihang University, Beijing, China
zhaow85@buaa.edu.cn
[2] The Beihang Hangzhou Innovation Institute, Yuhang, Xixi Octagon City,
Hangzhou, China

Abstract. Breast cancer ranks the first noncutaneous malignancy incidence and mortality in women worldwide, and seriously endangers the health and life of women. Ultrasound plays a key role and yet provides an economical solution for breast cancer screening. While valuable, ultrasound is still suffered from limited specificity, and its accuracy is highly related to the clinicians, resulting in inconsistent diagnosis. To address the challenge of limited specificity and inconsistent diagnosis, in this retrospective study, we first develop a learning model based on the computational ultrasound image features and identified a set of clinically relevant features. Then, the abstract spatial interaction patterns of the ultrasound images together with the extracted features were employed for breast malignancy diagnosis. We evaluate the proposed algorithm on the Breast Ultrasound Images Dataset (BUSI). The proposed algorithm achieved a diagnostic accuracy of 89.32% and a significant area under curve (AUC) of 0.9473 with the repeated cross-validation scheme. In conclusion, our algorithm shows superior performance over the existing classical methods and can be potentially applied to breast cancer screening.

Keywords: Computational features · Ultrasound · Breast cancer

1 Introduction

Breast cancer is the most commonly diagnosed cancer and causes the most deaths for women diagnosed with cancers. [6] Early diagnosis plays an important role in both treatment and prognosis for breast cancer. It has been extensively reported that patients diagnosed with smaller primary breast tumors had a significantly higher disease-free survival and overall survival, compared to patients with locally advanced breast tumors. Early detection and diagnosis of breast cancer are therefore of interest. Various imaging modalities have been applied to breast cancer diagnosis. Among these, ultrasound (US) imaging which employs sound waves to generate images of the internal morphology of the breast is the

W. Qin et al. (Eds.): CMMCA 2022, LNCS 13574, pp. 150–158, 2022.
https://doi.org/10.1007/978-3-031-17266-3_15

most widely used method due to its safety and painlessness. The US is able to help diagnose breast lumps and other abnormalities in a noninvasive way.

Despite its usefulness and wide applicability, breast US has suffered from limited specificity and interobserver variability, both of which contribute to a high rate of false-positive and false-negative. The misdiagnoses cause either a number of unnecessary biopsies and surgeries, or missed cases. To address the challenge of limited specificity and interobserver variability, there has been a growing interest in the application of machine learning technology for automatic US breast tumor identification [4].

Different from conventional US diagnosis, the machine learning approaches make decisions based on extracted computational features. The features extraction procedure can be performed using either deep neural networks [2] or spatial and texture computational tools. While the deep neural network-based features are usually illusive and lack interpretability, the spatial and texture computational tools extract features that are directly related to tumor size and shape, image intensity histogram, and relationships between image voxels from radiologic images. The mathematical definitions of these features are explicit and easy to reproduce. Some of these features, such as tumor texture, have been demonstrated to be useful for differentiating malignant from benign tumors in breast cancer. In this study, we aimed to develop a learning model based on the computational ultrasound image features and applied the model to breast tumor identification. Clinically relevant features were used to differentiate breast tumor malignancy.

2   Method

Radiomics researches have a rather clear pipeline [3] which we adopted. First, we prepared the data, where the segmentation of region of interest (ROI) had been already available. Next, we extract features from ROIs with PyRadiomics package. Then, we selected and eliminated features and prepared them for modeling. At last, we built our model and evaluated the model by common metrics. The adopted pipeline is shown in Fig. 1.

2.1   Data Preparation

The BUSI dataset [1] was collected from 600 female patients and divided into three categories: benign, malignant, and normal. Both ultrasound images and segmentation masks are stored as 8-bit pngs. A sample of a malignant ultrasound image, a benign ultrasound image, and their corresponding masks are shown in Fig. 2.

Since the radiomics extract information from the region of interest (ROI) instead of the entire image, an ultrasound image with more than one tumor will result in the situation that the number of tumors ROIs is greater than that of the ultrasound image. Through the pairing of the ultrasound images and the masks, 454 benign tumor ROIs and 211 malignant tumor ROIs were finally obtained.

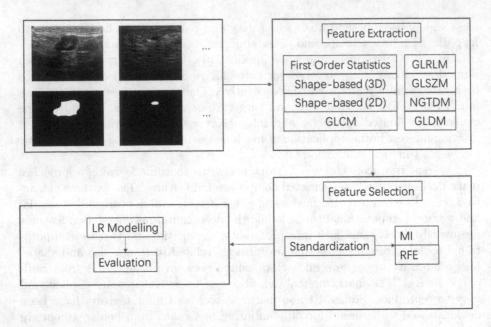

Fig. 1. Adopted pipeline of the research.

2.2   Feature Extraction

PyRadiomics [7] is an open-source Python library for radiomics feature extraction. With PyRadiomics, we extracted 1318 image-related features, which consist of eight classes:

- First Order Statistics
- Shape-based (2D)
- Shape-based (3D)
- Gray Level Cooccurence Matrix (GLCM)
- Gray Level Run Length Matrix (GLRLM)
- Gray Level Size Zone Matrix (GLSZM)
- Neigbouring Gray Tone Difference Matrix (NGTDM)
- Gray Level Dependence Matrix (GLDM).

2.3   Feature Selection

Features with too high dimension hinder the implementation of classification algorithms, so feature selection is required. After the following steps, the number of features is controlled in an appropriate range.

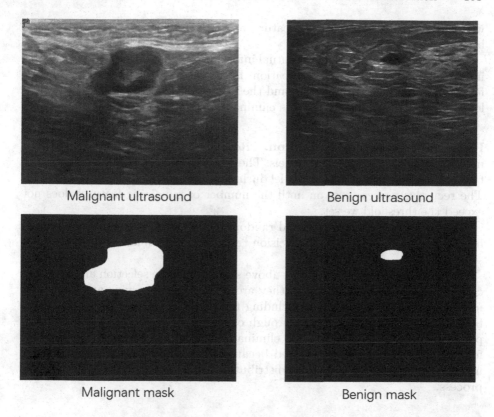

Fig. 2. Sample of a malignant ultrasound image, a benign ultrasound image, and their corresponding masks.

Data Standardization. The standardization process unifies the dimensions of the features and prevents the effect of the different magnitude order during the selection and modeling process.

We standardized the data by the formula

$$\hat{x} = \frac{x - \mu}{\sigma}, \tag{1}$$

where x represents the original data and $\hat{x}$ represents the standardized data. μ represents the mean of the data, and σ represents the standard deviation of the data.

Mutual Information Filtering The mutual information (MI) of a chosen feature X and label Y is defined as

$$I(X;Y) = E\left[I\left(x_i; y_j\right)\right] = \sum_{x_i \in X} \sum_{y_j \in Y} p\left(x_i, y_j\right) \log \frac{p\left(x_i, y_j\right)}{p\left(x_i\right) p\left(y_j\right)}, \tag{2}$$

where x_i represents the chosen feature of i-th sample, and y_j represents the binary label of j-th sample.

For a chosen feature, the less mutual information it has with the label, the less information it provides for classification. Based on this principle, we performed feature filtering based on the MI, and the features whose MI with the label was lower than the threshold of 0.1 was eliminated.

Recursive Feature Elimination. Recursive feature elimination (RFE) method works with predictive models. The feature which contributes the least to the result is determined by the model during each recursion and then eliminated. The recursive process goes on until the number of remaining features does not exceed the threshold we set.

In our implementation, we used random forest as the predictive model during the RFE process, where 25 decision trees were ensembled. 30 features were selected.

It is worth mentioning that the above steps of feature selection are not quite clear at the initial stage. Instead, they are determined by trying applying common feature selection methods(including filters, wrappers and embedded ones) by following the principles that through one single selection process, an appropriate number of features can be eliminated. Removing too many or too few features in one process are avoided because the extreme threshold of the former extremizes the training data distribution, and the latter fails the selection process.

2.4 Modeling and Evaluation

We chose linear regression, a simple machine learning model for the purpose of classification, with L_1 norm as the penalty, and liblinear as the solver. The max iteration was set to 10^4.

For evaluation, we used common metrics, including:

- F1-score
- Accuracy
- Sensitivity
- Specificity
- Precision
- ROC curve [5] and area under curve (AUC).

Each metrics were calculated with respect to the 30% test data for 50 random splits of the dataset.

3   Results and Discussions

3.1   Metrics Performance

The performance of the LR model on the selected metrics is listed in Table 1, and visualized in Fig. 3. The error bar indicates the 95% confidence interval (95% CI).

It can be seen from the figure that the model is robust to different split of training and test sets. Thus the metrics have a small interval of 95% CI.

The sensitivity is relatively low compared with other metrics. As the BUSI dataset suffers from data imbalance, where the number of available benign ROIs is nearly twice as that of malignant ones. Considering the definition of the sensitivity metric, it may be improved by properly oversampling the positive samples, i.e. malignant ROIs.

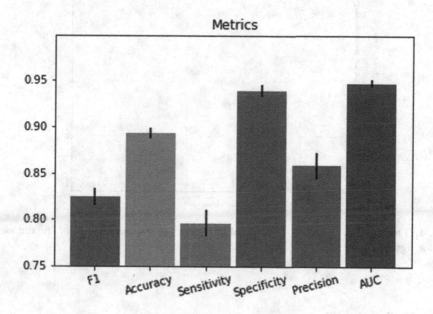

Fig. 3. Numerical metrics

Table 1. Metrics

F1-score	Accuracy	Sensitivity	Specificity	Precision	AUC
0.8246 ± 0.0089	0.8932 ± 0.0053	0.797 ± 0.014	0.9389 ± 0.0063	0.8584 ± 0.014.	0.9473 ± 0.004

3.2 ROC Curve

The ROC curve of the model on a random split of the dataset is shown in Fig. 4. The corresponding AUC is 0.9469.

The ROC curve and corresponding AUC reveal that the model has a relatively high predictive value from an overall perspective, especially considering the imbalance of the dataset in this study.

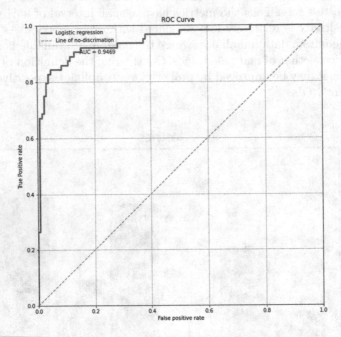

Fig. 4. The blue line is the ROC curve of our model on a random split of the dataset. (Color figure online)

3.3 Calibration Curve

The calibration curve corresponding to the model with the ROC Curve above is shown below in Fig. 5. As can be seen from the figure, when the predicted value is at lower (<0.3) and higher (>0.7) values, the calibration curve of the model is close to the perfectly calibrated curve. The deviation on the interval around 0.5 indicates that the model has much room for improvement. Attaching attention technologies or simply put more weight on the training samples whose predicted value falls in the interval around 0.5 may lead to the calibration curve approaching to the perfectly calibrated one and improve the performance of the model.

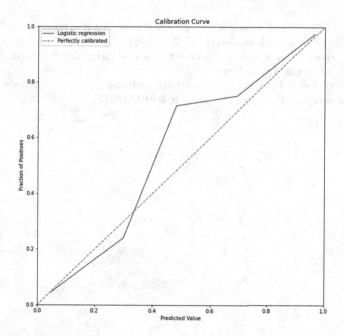

Fig. 5. The calibration curve of a random split.

4 Conclusion

We present a computational US image modeling algorithm to accurately identify breast tumors. The algorithm is able to extract reproducible and interpretable features to differentiate breast tumor malignancy. Using these clinically relevant features, the proposed classification model achieves promising results based on clinical US images from public BUSI dataset. We anticipate that the proposed tumor identification and feature extraction and selection scheme can adapt to a broader category of cancers.

Acknowledgements. This work was supported in part by the National Natural Science Foundation of China (No. 12175012).

References

1. Al-Dhabyani, W., Gomaa, M., Khaled, H., Fahmy, A.: Dataset of breast ultrasound images. Data Brief **28**, 104863 (2020)
2. Becker, A.S., Mueller, M., Stoffel, E., Marcon, M., Ghafoor, S., Boss, A.: Classification of breast cancer in ultrasound imaging using a generic deep learning analysis software: a pilot study. Br. J. Radiol. **91**(1083), 20170576 (2018)
3. Bibault, J.E., et al.: Radiomics: a primer for the radiation oncologist. Cancer/Radiothérapie **24**(5), 403–410 (2020)
4. Cole-Beuglet, C., Beique, R.A.: Continuous ultrasound B-scanning of palpable breast masses. Radiology **117**(1), 123–128 (1975)

5. Cook, N.R.: Statistical evaluation of prognostic versus diagnostic models: beyond the ROC curve. Clin. Chem. **54**(1), 17–23 (2008)
6. Ferlay, J., et al.: Global cancer observatory: cancer today. International Agency for Research on Cancer, Lyon, France (2020)
7. Van Griethuysen, J.J., et al.: Computational radiomics system to decode the radiographic phenotype. Can. Res. **77**(21), e104–e107 (2017)

Author Index

Printed in the United States
by Baker & Taylor Publisher Services